AF845302

Physical Modeling of Mixing in Water Storage Tanks

Awwa Research Foundation

About the Awwa Research Foundation

The Awwa Research Foundation (AwwaRF) is a member-supported, international, nonprofit organization that sponsors research to enable water utilities, public health agencies, and other professionals to provide safe and affordable drinking water to consumers.

The Foundation's mission is to advance the science of water to improve the quality of life. To achieve this mission, the Foundation sponsors studies on all aspects of drinking water, including supply and resources, treatment, monitoring and analysis, distribution, management, and health effects. Funding for research is provided primarily by subscription payments from approximately 1,000 utilities, consulting firms, and manufacturers in North America and abroad. Additional funding comes from collaborative partnerships with other national and international organizations, allowing for resources to be leveraged, expertise to be shared, and broad-based knowledge to be developed and disseminated. Government funding serves as a third source of research dollars.

From its headquarters in Denver, Colorado, the Foundation's staff directs and supports the efforts of more than 800 volunteers who serve on the board of trustees and various committees. These volunteers represent many facets of the water industry, and contribute their expertise to select and monitor research studies that benefit the entire drinking water community.

The results of research are disseminated through a number of channels, including reports, the Web site, conferences, and periodicals.

For subscribers, the Foundation serves as a cooperative program in which water suppliers unite to pool their resources. By applying Foundation research findings, these water suppliers can save substantial costs and stay on the leading edge of drinking water science and technology. Since its inception, AwwaRF has supplied the water community with more than $300 million in applied research.

More information about the Foundation and how to become a subscriber is available on the Web at **www.awwarf.org.**

Physical Modeling of Mixing in Water Storage Tanks

Prepared by:
Philip J. W. Roberts, **Xiaodong Tian**, and **Fotis Sotiropoulos**
School of Civil and Environmental Engineering
Georgia Institute of Technology, Atlanta, GA 30332
and

Michael Duer
Tideflex Technologies, Inc.
300 Bilmar Drive, Pittsburgh, PA 15205

Sponsored by:
Awwa Research Foundation
6666 West Quincy Avenue, Denver, CO 80235-3098

Published by:

Awwa Research Foundation

American Water Works Association

IWA Publishing

DISCLAIMER

This study was funded by the Awwa Research Foundation (AwwaRF). AwwaRF assumes no responsibility for the content of the research study reported in this publication or for the opinions or statements of fact expressed in the report. The mention of trade names for commercial products does not represent or imply the approval or endorsement of AwwaRF. This report is presented solely for informational purposes.

CONTENTS

LIST OF TABLES

LIST OF FIGURES

FOREWORD

The Awwa Research Foundation is a nonprofit corporation that is dedicated to the implementation of a research effort to help utilities respond to regulatory requirements and traditional high-priority concerns of the industry. The research agenda is developed through a process of consultation with subscribers and drinking water professionals. Under the umbrella of a Strategic Research Plan, the Research Advisory Council prioritizes the suggested projects based upon current and future needs, applicability, and past work; the recommendations are forwarded to the Board of Trustees for final selection. The foundation also sponsors research projects through the unsolicited proposal process; the Collaborative Research, Research Applications, and Tailored Collaboration programs; and various joint research efforts with organizations such as the U.S. Environmental Protection Agency, the U.S. Bureau of Reclamation, and the Association of California Water Agencies.

This publication is a result of one of these sponsored studies, and it is hoped that its findings will be applied in communities throughout the world. The following report serves not only as a means of communicating the results of the water industry's centralized research program but also as a tool to enlist the further support of the nonmember utilities and individuals.

Projects are managed closely from their inception to the final report by the foundation's staff and large cadre of volunteers who willingly contribute their time and expertise. The foundation serves a planning and management function and awards contracts to other institutions such as water utilities, universities, and engineering films. The funding for this research effort comes primarily from the Subscription Program, through which water utilities subscribe to the research program and make an annual payment proportionate to the volume of water they deliver and consultants and manufacturers subscribe based on their annual billings. The program offers a cost-effective and fair method for funding research in the public interest.

A broad spectrum of water supply issues is addressed by the foundation's research agenda: resources, treatment and operations, distribution and storage, water quality and analysis, toxicology, economics, and management. The ultimate purpose of the coordinated effort is to assist water suppliers to provide the highest possible quality of water economically and reliably. The true benefits are realized when the results are implemented at the utility level. The foundation's trustees are pleased to offer this publication as a contribution toward toward that end.

Walter J. Bishop
Chair, Board of Trustees
Awwa Research Foundation

Robert C. Renner, P.E.
Executive Director
Awwa Research Foundation

ACKNOWLEDGMENTS

The authors are most appreciative of Mr. Andy Udell for his expert construction, usually on short notice, of the many configurations tested, and for his other help in the lab without which this study would not have been possible.

We are also grateful to the members of the review panel: Walter Grayman, Lew Rossman, and Cree Horner for their very helpful comments on the early work and proposed plans and for their continuing interest and support of this project.

EXECUTIVE SUMMARY

The results of extensive experimental studies on jet-induced mixing in water storage tanks are reported. The experiments were conducted using a newly developed three-dimensional laser-induced fluorescence (3DLIF) system that can measure the whole field of tracer concentrations in the tanks and its temporal evolution through the mixing process. Experiments were conducted on three types of storage tanks: Ground level cylindrical (GC), Standpipe (ST), and Ground level rectangular (GR). For each tank style, various combinations of inlet geometries were tested. The inlet geometries included single and multiple nozzles that were oriented vertically, horizontally, or at 45° to the vertical. The inlets were at the tank bottoms or distributed over their depths. Various flow rates were tested with inflowing water that was either neutrally, positively, or negatively buoyant compared to the water in the tank. A total of about 140 experiments were done, forming a huge data base that should aid in rational tank design and in the development of computational fluid dynamics (CFD) codes.

For each experiment, the distribution of tracer concentration within the tank was measured as a function of time. The degree of mixing was quantified by the coefficient of variation (COV) of the tracer concentration distribution, defined as the standard deviation of the concentrations in the tank divided by their mean value. A time for mixing t_m was defined as the time where the COV fell to 10%. A dimensionless time for mixing, τ_m, was defined based on t_m, the tank volume, and the momentum flux of the inflow. The dimensionless time was used to compare the mixing efficiencies of the various tank and inlet configurations.

The experiments reveal complex flow patterns within the tanks, even in the absence of buoyancy effects. Three-dimensional circulations and gyres can form. In particular, tanks with a vertical center nozzle with large diameter-to-depth ratio formed a ring-shaped dead zone. This flow structure has not been previously reported. Moving the nozzle to a side wall also resulted in formation of a recirculation gyre, but it was not a dead zone that prevented mixing.

Values of dimensionless mixing times are presented from which mixing times in prototype tanks can be predicted.

The following observations were made for experiments with no density differences and for runs over one inflow cycle with inflow only.

- For cylindrical tanks:
  - The dimensionless mixing time $\tau_m \approx 9$ for ground level cylindrical tanks with $H/D \approx 0.25$.
  - The mixing time increases as the depth-to-diameter ratio increases. For standpipe tanks $\tau_m \approx 15$ for $H/D \approx 2.5$. In other words, tall slender tanks are more difficult to mix than short, squat tanks.
  - Vertical nozzles at the bottom mixed most efficiently. Mixing became more rapid as the number of nozzles increased. For GC tanks, four nozzles distributed over the tank bottom reduced the mixing time by about 50% compared to a single vertical nozzle. Multiple horizontal nozzles of nozzles oriented at 45° did not decrease mixing time.
  - If a single vertical nozzle is used, placing it near a sidewall appears to be preferred.
  - The use of a draft tube in tall standpipe tanks impairs mixing due to restriction of entrainment into the jet.
- Mixing times for ground level rectangular (GR) tanks are generally similar to GC tanks.

Density differences (usually caused by temperature differences between the inflow and the stored water) can complicate the flow considerably. Whether or not the tank mixes depends mainly on the density difference, the inflow momentum flux, and the water depth. If the tank does not mix, a stable density stratification forms which could be quite resistant to subsequent mixing. The following observations were made.

- For negatively buoyant inflows (i.e. the inflow is colder than the stored water):
  - Vertical discharges at the bottom are best.
  - The tendency to mix depends on the total momentum flux and is only weakly improved by multiple nozzles. A criterion is given to predict whether mixing occurs with vertical or inclined nozzles based on the vertical component of the total momentum flux, the buoyancy flux, and the water depth. This criterion applied to all tank styles.
  - Of the tanks with horizontal nozzles, only those with high momentum flux with multiple nozzles distributed over the water depth became mixed.
  - A draft tube improved mixing for some standpipe tests. This is because the standpipe provided a route for the dense inflow to reach the top of the tank before falling and mixing. Some of the tanks became mixed with the draft tube that would otherwise not be mixed.
- For positively buoyant inflows (i.e. the inflow is warmer than the stored water. These tests were only with GC tanks):
  - Horizontal inflows at the bottom mixed best.
  - None of the single vertical nozzles resulted in mixing.
  - Criterion for mixing were given based on the momentum flux, the density difference, and the water depth.

Some tests were made on the GC tank with simultaneous inflow and outflow at the same rates so the volume in the tank remained constant. The mixing times were longer than for similar conditions with inflow only.

The flows are dominated by large-scale, coherent unsteady vortices, which account for a significant part of transport and mixing. CFD models of tank mixing will have to simulate such structures and the complex three-dimensional circulations that were observed in order to be reliable tools to predict tank mixing and for tank design. 3DLIF is a cost-effective tool for investigating and comparing tank designs to ensure efficient mixing.

CHAPTER 1.
INTRODUCTION

Deterioration of water quality in finished water storage facilities is a major concern for water utilities. Despite extensive investigations over a number of years, many problems still plague the estimated 400,000 finished water storage facilities in the United States. These problems are described in a recent issue of the AWWA Journal (Martel et al., 2002). They include loss of disinfectant residual due to hydraulic short-circuiting, poor mixing and circulation, poor turnover time, and excessive detention time. Reservoirs, with their large capacities and long detention times, are generally considered to have negative impacts on water quality within the distribution system that can often be attributed to the poor locations, configurations, and orientations of inlet and outlet pipes.

The mixing processes in these storage tanks are primarily driven by the momentum of the inflow. A complex flow results that is three-dimensional, unsteady, and difficult to predict. The difficulties are compounded by the enclosing geometry and also buoyancy forces caused by density differences due to temperature differences between the inflow and the stored water. As a result, there are few guidelines on how to design storage tanks to promote effective mixing. Although laboratory and field studies on mixing tanks have been done, they have been limited by point sampling techniques to observations at only a few points in the tanks that are inadequate to portray the three-dimensionality of the flows.

In this report, we present the results of research on tank mixing using a three-dimensional laser-induced fluorescence (3DLIF) system that allows non-intrusive measurements of the entire tracer concentration field in the tank. Systematic experiments were conducted on three major reservoir types with varying inlet and outlet geometries, including multiple nozzle diffusers, that include many tank designs found in practice. Various inlet geometries, inflow rates, and density differences were modeled. The experiments take into account jet momentum and also buoyancy effects that may be caused by differences between the inlet water and tank water temperatures. The range of buoyancies studied was representative of those that commonly occur due to temperature variations around the US and worldwide. The overall objective of the project was to obtain basic information on the hydrodynamics of jet-induced mixing in closed tanks and to provide recommendations for the design of storage tanks that enhance water quality by maximizing mixing.

About 140 experiments were conducted and extensive data were collected and analyzed. In this report, the experimental program is described and the results of experiments presented. Some CFD simulations were also done to elucidate the issues facing numerical modeling of these complex flows and to indicate future research needs in developing reliable CFD tools. In addition to this report, animations of some of the experiments can be seen on the accompanying CD.

CHAPTER 2.
EXPERIMENTAL APPARATUS AND TECHNIQUES

CONFIGURATIONS TESTED

The objective of this research was to investigate jet-induced mixing in water storage tanks. Laboratory experiments were performed in order to measure the influence of tank geometry and inlet configurations for varying inflow momentum with and without density differences between the inflow and the stored water. Three common styles of distribution storage tanks were tested as shown in Figure 2.1.

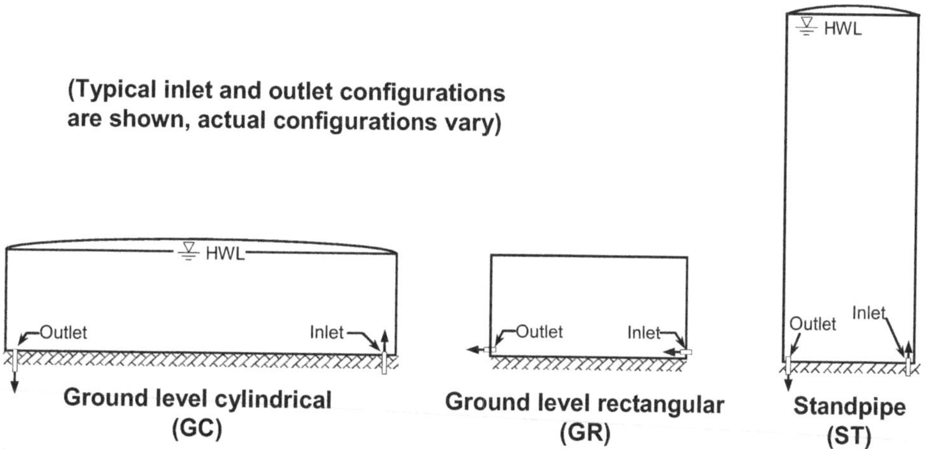

Figure 2.1 Tank styles tested

The tank styles are: Ground level cylindrical (GC); Ground level rectangular (GR); and Standpipe (ST). Typical tank dimensions are shown in Table 2.1, along with the dimensions of the models used. Twenty six combinations of tank and inlet/outlet configurations, shown schematically in Table 2.2, were tested. The physical modeling laws are discussed in Chapter 2. The results can be scaled to other geometrically similar tanks of various sizes, inflow rates, and densities using the scaling methods discussed in Appendix A. The conditions and results for all experiments are summarized in Appendix B.

Table 2.1 Dimensions of Prototype and Model Tanks.

Tank style	Prototype				Model			
	Diameter or sidewall length (ft)	Water depth (ft)	Nozzle diameter (in)	Volume (MG)	Scale	Diameter or sidewall length (in)	Water depth (in)	Nozzle diameter (in)
Ground level cylindrical (GC)	119.0	30.0	16.0	2.5	81.6	17.5	4.4	0.196
Standpipe (ST)	50.0	125.0	16.0	1.8	80.0	7.5	18.8	0.200
Ground level rectangular (GR)	105.0	30.0	16.0	2.5	81.6	15.4	4.4	0.196

Table 2.2 Tank and Nozzle Configurations Tested.

Ground level cylindrical (GC)	Standpipe (ST)	Ground level rectangular (GR)
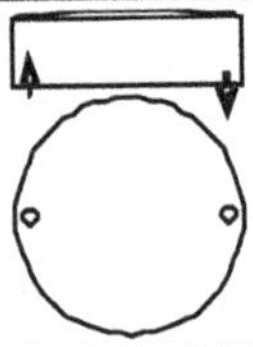 GC01: One nozzle, vertical, near wall. Outlet near wall for simultaneous inflow and outflow experiments.	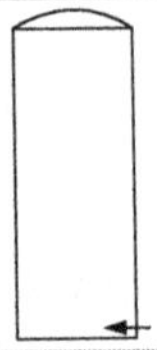 ST01: One nozzle, horizontal, near floor	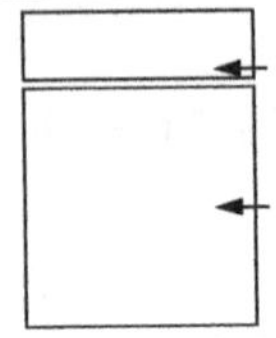 GR01: One nozzle, horizontal, near floor, mid-wall
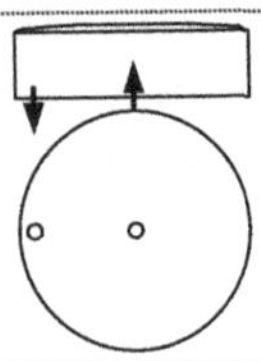 GC02: One nozzle, vertical, center Outlet near wall for simultaneous inflow and outflow experiments.	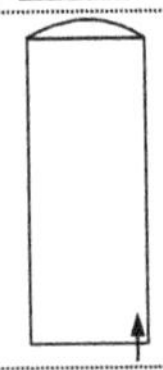 ST02: One nozzle, vertical, near wall	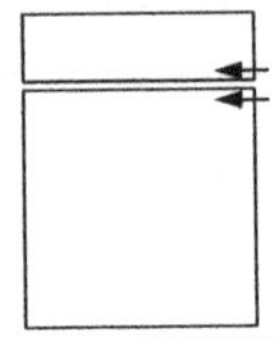 GR02: One nozzle, horizontal, near floor, near side wall
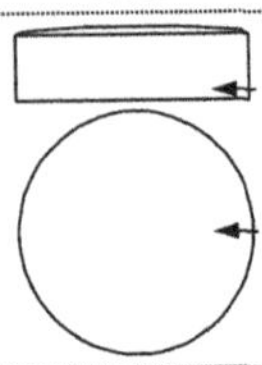 GC03: One nozzle, horizontal, near floor	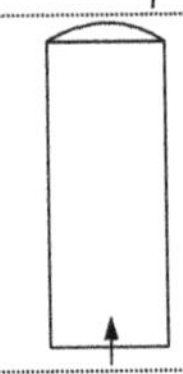 ST03: One nozzle, vertical, center	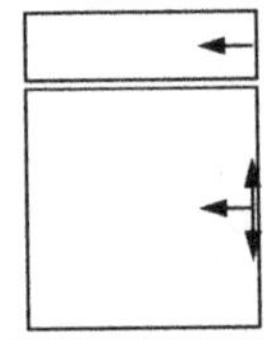 GR03: Three nozzles, horizontal, mid-depth
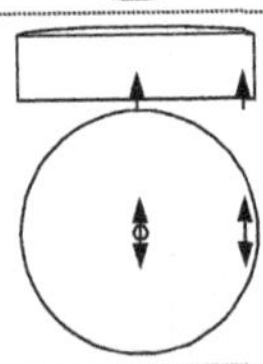 GC04: Five nozzles. Two near wall up at 45°. Three at center, one vertical, two at 45°	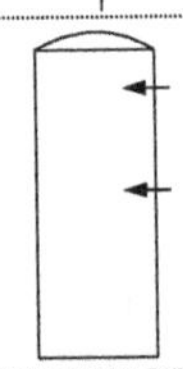 ST07: Two nozzles, horizontal mid-depth and near surface	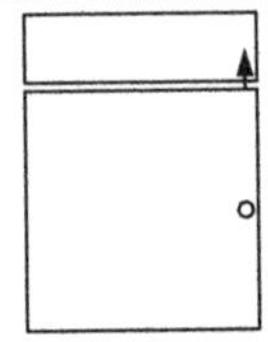 GR04: One nozzle, vertical
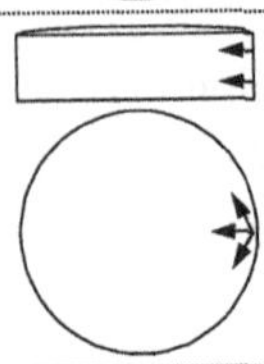 GC05: Three nozzles, horizontal. One at 2/3 depth, two at 1/3 depth tangential to wall	 ST09: Seven nozzles, horizontal, distributed vertically	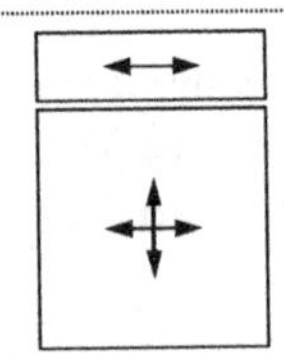 GR05: Four nozzles, horizontal, mid-depth
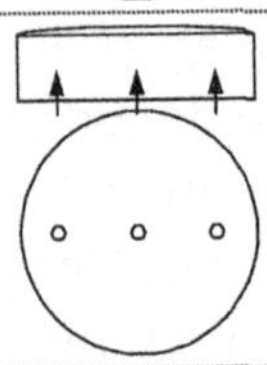 GC06: Three nozzles, vertical	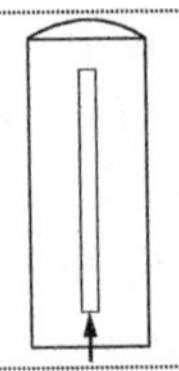 ST11: One nozzle, vertical, center, plus draft tube	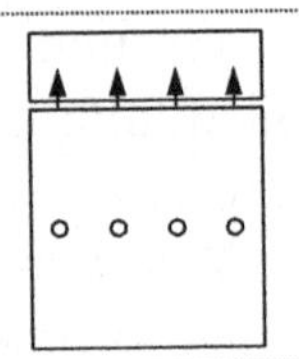 GR06: Four nozzles, vertical
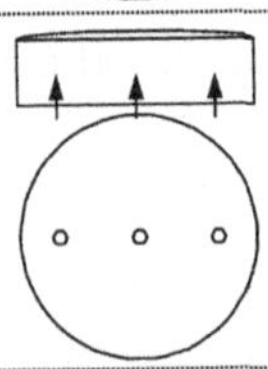 GC07: Six nozzles, up at 45°. Same as GC06, except two nozzles per riser.	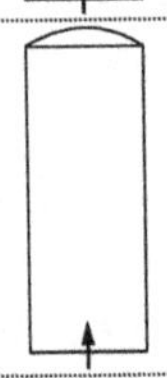 ST12: Same as ST11 but no draft tube	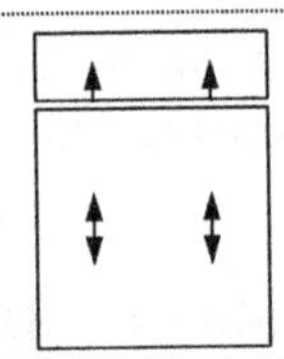 GR07: Four nozzles, inclined up at 45°
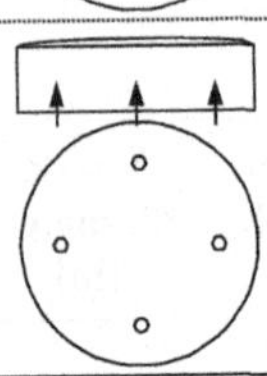 GC08: Four nozzles, vertical, one in each quadrant		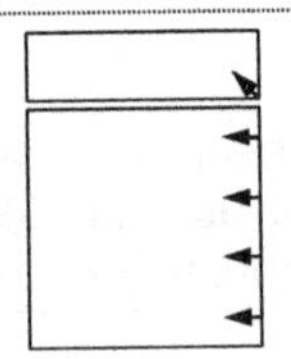 GR08: Four nozzles, inclined up at 45°
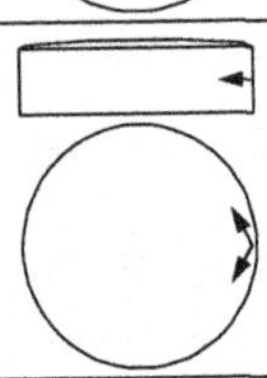 GC09: Two nozzles, mid-depth, tangential		

The experiments account for the momentum of the inflowing jets and also buoyancy effects due to density differences between the inflow and the stored water. The density differences modeled are based on expected ranges that occur in different seasons and geographic regions. Most of the experiments were conducted with inflow only; a few were done with simultaneous inflow and outflow (flow-through).

Some inlet configurations were tested for two inflow rates (high and low momentum fluxes) and three buoyancies: neutral, positive, and negative. Not every possible combination of variables could be tested, so each tank style was first run at two flowrates with neutral buoyancy to establish the effects of momentum on mixing. Experiments were then run at the lower flow rate with positive and negative buoyancies.

PHYSICAL MODELING LAWS

Hydraulic models are usually scaled according to laws based on traditional dimensionless parameters, for example, Reynolds number, Froude number, and length scales. It is never possible to ensure full similitude in a hydraulic model; in other words it is never possible to have simultaneous equality of all dimensionless parameters. Therefore, it is essential that the dominant processes and forces are modeled.

The flows considered here are driven primarily by the jet-induced momentum of the inflow but may also be influenced by density differences between the inflow and stored water. Such an inflow is generally described as a buoyant jet. It is characterized by the jet velocity u_j, the nozzle diameter d, and the modified acceleration due to gravity, $g_o' = g(\Delta\rho/\rho_a)$ (provided $\Delta\rho/\rho_a << 1$, which it always will be here) where $\Delta\rho = \rho_a - \rho_o$ is the density difference between the stored water, ρ_a, and the inflow, ρ_o.

Consider a round tank, characterized by its diameter D and water depth H, that is mixed by the inflow of a buoyant jet. Any flow property, such as the time required for the tank to be fully mixed, t_m, is given by:

$$t_m = f(u_j, d, g_o', D, H) \tag{2.1}$$

which becomes, following a dimensional analysis:

$$\frac{t_m u_j}{D} = f\left(\frac{H}{D}, \frac{d}{D}, F_j\right) \tag{2.2}$$

where $F_j = u_j \big/ \sqrt{g_o' d}$ is the jet densimetric Froude number. The modeling laws would then consist of equality of the Froude number and of the length scale ratios, H/D and d/D, between model and prototype (full-scale). This is referred to as an undistorted model, i.e. it is geometrically similar to the prototype as all lengths scale in the same ratio. These laws are often used in modeling the behavior of buoyant jets that arise in wastewater discharges into the environment, for example Roberts and Snyder (1993). Equality of the jet Reynolds number, $\mathrm{Re} = u_j d/\nu$ where ν is the kinematic viscosity, between model and prototype is not possible, and Re is much smaller in the model than in the prototype. This is not a serious limitation, however, as the flow properties are independent, or only weakly dependent on Re (and of viscous effects) provided the jets are fully turbulent. This occurs if Re > 2000 (Fischer et al., 1979, Roberts and Snyder, 1993).

The number of variables can be reduced, and greater physical insight obtained, by characterizing the jet by its fluxes of volume, *Q*, momentum, *M*, and buoyancy, *B*:

$$Q = \frac{\pi}{4} d_j^2 u_j \qquad M = u_j Q \qquad B = g_o' Q \tag{2.3}$$

Because the dynamical effect of *Q* can usually be neglected (except very close to the nozzle) the most important parameters are *M* and *B*. Eq. 2.1 can then be replaced by:

$$t_m = f(M, B, D, H) \tag{2.4}$$

which becomes, following a dimensional analysis:

$$\frac{t_m M^{1/2}}{D^2} = f\left(\frac{H}{D}, \frac{l_M}{H}\right) \tag{2.5}$$

where $l_M = M^{3/4}/B^{1/2}$ is a length-scale of the buoyant jet. Similitude now results from equality of the ratio *H/D* (and of course the tank shape and inlet location and orientation) and the ratio l_M/H. It is no longer necessary to geometrically scale the nozzle diameter so long as the ratio l_M/H is the same in model and prototype. This allows for wider application of the scaling results. Note that $l_M/H = (\pi/4)^{1/4} F_j/(H/d)$.

The ratio l_M/H has dynamical significance. For a freely rising plume, l_M is the distance over which the jet momentum is important relative to the jet buoyancy flux (Fischer et al., 1979). If $l_M/H \ll 1$, the flow is buoyancy dominated and the source momentum has negligible effect; if $l_M/H \gg 1$ the flow is momentum dominated and the buoyancy has negligible effect. The flow of a vertical buoyant jet rising upwards in water of finite depth *H* becomes unstable (mixes over the depth) if $l_M/H > 0.23$, but forms a stable, stratified surface flow if $l_M/H < 0.23$ (Wright et al., 1991). While stable flow is desirable for a wastewater discharge into the environment, as it prevents reentrainment of the inflow back into the plume, it is undesirable in mixing tanks where it results in a stable stratification forming in the tank that inhibits mixing. These considerations allow better classification of the flow types possible, and better choices of conditions to model.

A special case occurs when where there is no density difference between the inflow and the stored water. *B* is then zero, and Eq. 2.5 becomes:

$$\frac{t_m M^{1/2}}{D^2} = f\left(\frac{H}{D}\right) \tag{2.6}$$

For a fixed value of *H/D*, $t_m M^{1/2}/D^2$ is equal to a constant, an equation proposed by Fossett and Prosser (1949), and Van de Vusse (1955). Because the tank volume, *V,* is proportional to HD^2, Eq. 2.6 can also be written:

$$\tau_m = \frac{t_m M^{1/2}}{V^{2/3}} = f\left(\frac{H}{D}\right) \tag{2.7}$$

where $\tau_m = t_m M^{1/2}/V^{2/3}$ is a dimensionless mixing time. For fixed *H/D*, Eq. 2.7 becomes:

$$\tau_m = \frac{t_m M^{1/2}}{V^{2/3}} = \text{Constant} \tag{2.8}$$

as proposed by Rossman and Grayman (1999). Eq. 2.7 shows, however, that it is only valid over some limited range of *H/D*; an objective of this research is to investigate the conditions for which Eq. 2.8 applies.

The requirements for similitude between model and prototype (i.e. the modeling laws) for a flow with density differences follow from Eq. 2.5:

$$\left(\frac{l_M}{H}\right)_p = \left(\frac{l_M}{H}\right)_m \quad \text{and} \quad \left(\frac{H}{D}\right)_p = \left(\frac{H}{D}\right)_m \tag{2.9}$$

or

$$\left(\frac{l_M}{H}\right)_r = 1 \quad \text{and} \quad \left(\frac{H}{D}\right)_r = 1 \tag{2.10}$$

where the subscript *p* denotes the prototype (full-scale), *m* denotes the model, and *r* the ratio of prototype to model.

There are three independent choices of model scales: The density difference ratio, $(\Delta\rho/\rho)_r$, the geometric scale, D_r (also equal to H_r), and the nozzle diameter ratio d_r. Once these scales are chosen, all other ratios are automatically determined. For example, the flow rate ratio Q_r ($= Q_p/Q_m$) is:

$$Q_r = \left(\frac{\Delta\rho}{\rho}\right)_r^{1/2} D_r d_r^{3/2} \tag{2.11}$$

the time ratio, t_r ($= t_p/t_m$) is:

$$t_r = \left(\frac{\Delta\rho}{\rho}\right)_r^{1/2} D_r^{-2} d_r^{3/2} \tag{2.12}$$

and the jet velocity ratio u_{jr} ($= u_{jp}/u_{jm}$) is:

$$u_{jr} = \left(\frac{\Delta\rho}{\rho}\right)_r^{1/2} D_r d_r^{-1/2} \tag{2.13}$$

Note that $D_r = H_r$, so the model tank diameter and depth are undistorted, but d_r need not be equal to D_r, i.e. the nozzle diameters need not be geometrically scaled in the same ratio. Also, the density difference ratio $(\Delta\rho/\rho)_r$ need not be equal to one, i.e. density differences in the model need not be the same as in the prototype. For further discussions of similar modeling, see Roberts and Snyder (1993).

The case with no density difference, i.e. $B = 0$ so that the inflow is characterized only by the momentum flux M, is an interesting one. The flow characteristics now depend only on M, H, and D, implying that equality of one dimensionless group is required for similitude. But no dimensionless group can be formed from M, H, and D, that includes M; in other words, the results of model tests with no density differences can be scaled to any (geometrically similar) tank of any size and any inflow rate. This is the same statement as Eq. 2.8, but arrived at from a different point of view.

For inflows with density differences, the results can be scaled to tanks of other sizes, inflow, and density difference by making the independent dimensionless parameters of Eq. 2.5 equal in model and prototype. For inflows with no density difference, Eq. 2.8 is applicable, using the appropriate value of the constant for the particular tank geometry. The prototype tanks should have similar nozzle geometry, i.e. the number of nozzles, their orientation, and location, and the Reynolds numbers of the inlets should be greater than about 2,000 to ensure turbulent flow (this will almost always be the case for large prototype tanks). Examples of scaling the results to various prototype conditions are given in Appendix A.

APPARATUS

A newly developed three-dimensional laser induced fluorescence (3DLIF) system was used to measure the evolution of the mixing induced by the inflow and the spatial variation of tracer concentration. The experiments were conducted in the Environmental Fluid Mechanics Laboratory at the Georgia Institute of Technology.

The experimental configuration is shown in Figure 2.2. The test tanks are constructed of clear Lucite to allow the laser beam to pass through. Two of the tank styles are cylindrical and their curved sidewalls would cause refraction of the laser beam as it passed through the air/Lucite and Lucite/water interfaces. To obviate this, the test tanks are placed in a large rectangular tank which is filled with water. The outer tank is about 1.0 m wide by 0.86 m deep by 0.61 m tall. The left and rear panels and tank floor are painted black to reduce reflections. A photograph of the system in operation is shown in Figure 2.3.

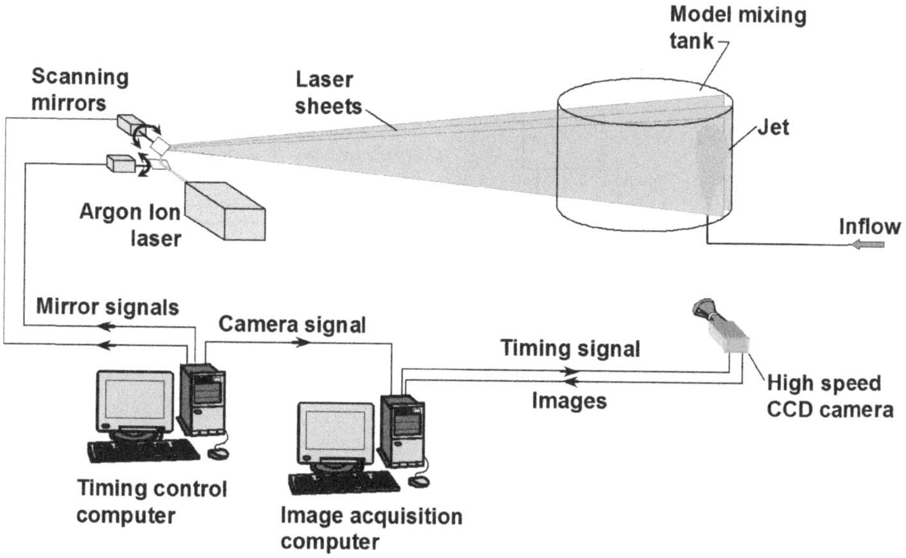

Figure 2.2 Schematic depiction of 3DLIF system

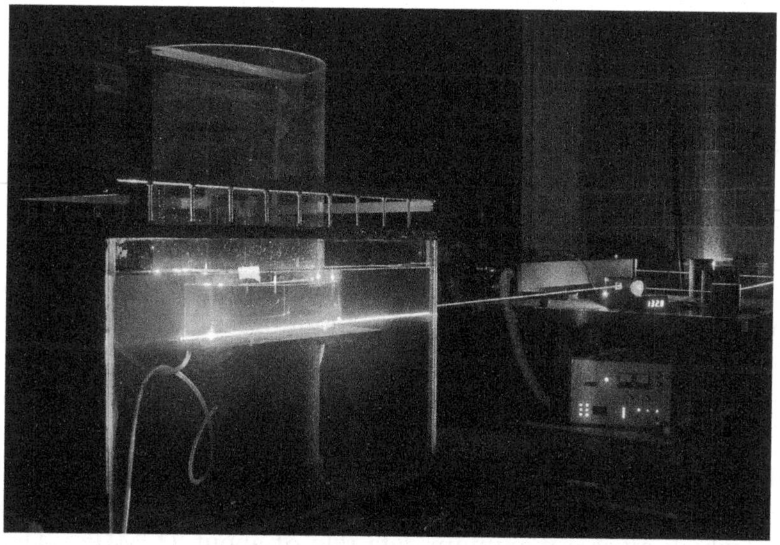

Figure 2.3 Photograph of experiment

The inflow is supplied by the system shown in Figure 2.4. The solution, stored in a 40-liter tank, is a mixture of fresh water and fluorescent dye with salt (NaCl) and/or ethanol added for density control. Sodium thiosulfate is added to remove any residual chlorine. The fluid densities are measured with a calibrated Troemner specific gravity balance to an accuracy of $\pm 0.1\ \sigma_t$ (± 0.1 kg/m^3). The inflow is pumped from the source tank and the flowrate is measured by a turbine flowmeter whose analog voltage output is monitored by a laptop computer containing a data acquisition board. This computer can control the flowrate by means of a proportioning solenoid valve.

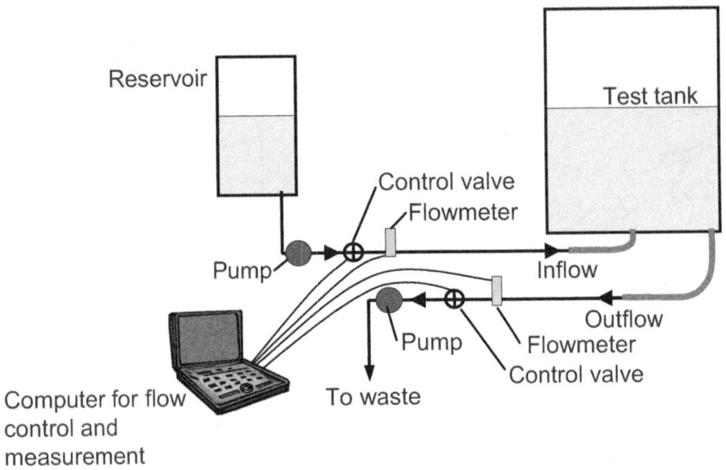

Figure 2.4 Schematic depiction of experimental flow control

The LIF system consists of two fast scanning mirrors that drive a laser beam from an Argon-Ion laser through the flow in a programmed pattern (Figure 2.2). The system is controlled by two computers, one for overall timing control, and one for image capture. A small amount of a fluorescent dye is added to the inflow. The fluorescent dye is Rhodamine 6G, whose wavelength of maximum absorption is close to that of the laser (514 nm). Detailed characteristics of Rhodamine 6G are given in Ferrier et al. (1993). The laser causes the dye to fluoresce, and the emitted light is captured by a CCD camera. The vertical mirror sweeps the beam down and back while the camera is exposing (i.e. the shutter is "open"). The horizontal mirror then moves the beam a small distance horizontally, the previous frame is downloaded, the camera buffer cleared, and the next exposure begins. This is repeated so that multiple vertical "slices" through the flow are obtained. After a predetermined number of "slices" the beam returns to the starting point and the cycle starts again. The images are written to hard disc in real time and saved for further processing. Tracer concentrations are obtained from the images by the methods discussed below.

The camera is a Dalsa CA-D6. This digital camera provides output in 8-bit resolution, i.e., a gray scale with 256 levels. The resolution (number of active pixels) is 532 by 516. The LVDS (Low Voltage Differential Signal, also known as EIA-644) data format enables high data transmission rates over long cable lengths. The maximum frame rate of this camera is 260 frames per second, which gives a maximum data rate of about 71 MHz. This high data rate is achieved by using four taps, each capable of 25 MHz. For the experiments reported here, frame rates of 100 frames per second were used. The camera is externally triggered by a TTL signal from the National Instruments I/O Board. It has a high-gain A/D converter to enable use with low fluorescence light levels. Even with the high gain, the noise level is still quite low. A Fujinon CCTV camera lens of 25 mm focal length and f0.85 aperture was used. A long pass orange filter (Schott glass 530) was placed between the lens and the CCD sensor lens to pass only the fluoresced light and eliminate the laser scattered light.

CONFIGURATIONS AND MODEL SCALES

The parameters for all experiments are summarized in Appendix B. The experiments were designed according to the hydraulic modeling laws (Eqs. 2.9 and 2.10) assuming nominal scales of 80:1 or 81.6:1 and the prototype dimensions shown in Table 2.1. The results can be scaled to other prototype conditions, however, as discussed in Appendix A.

EXPERIMENTAL PROCEDURE

The test tank and outer rectangular tank (if used) were first filled with filtered tap water and allowed to settle. The inflow solution is then prepared and the fluid densities measured. The experiment is initiated by a digital signal from the timing control computer (Figure 2.2) which initiates image capture by the CCD camera. The inflow is then begun and the flowrate is monitored by the laptop computer. Images were obtained at a rate of 100 frames per second. To improve image quality, the laser beam was scanned four times vertically for each image, i.e. the laser scan frequency is 400 Hz. Each full 3D data set consists of 40 slices, or images. There is a short delay, typically 1.2 to 3 seconds, between every complete set of 3D scans. The inflow rate was maintained constant for a finite time, then stopped. During a typical experiment 4500 to 5700 images are obtained, requiring a storage space of 1 to 1.5 gigabytes. The raw data were backed up and archived to DVD.

The images were corrected for spatial lens and sensor variations using the procedures detailed in Daviero et al. (2001) and Tian (2002). A "standard" image is obtained of a uniform white board illuminated by a uniform light source. The gray scale levels of such an image would be uniform for a perfect lens and sensor. Typically, however, it is usually brighter at the center and darker at the edges, a phenomenon known as vignetting. This standard image is used to correct for this non-uniform brightness variation. A black level image is also obtained by covering the lens. This image is also non-uniform due to variations in pixel characteristics. These variations are corrected as follows. The black level image is first subtracted from each raw image and also from the standard image. The new raw images are then divided by the new standard image, pixel by pixel, and the final images are then multiplied by a scale factor:

$$I_c(i,k) = K \times \frac{I_r(i,k) - I_b(i,k)}{I_s(i,k) - I_b(i,k)} \tag{2.14}$$

where (i,k) are the pixel indices, $I_c(i,k)$ is the corrected pixel value, $I_r(i,k)$ is the raw image pixel value, $I_s(i,k)$ is the standard image pixel value, $I_b(i,k)$ is the black level image pixel value, and K is a scale factor equal to the average pixel value in the standard image.

The laser beam decreases in power as it traverses the water in the tanks causing decreased light emission. This attenuation can be substantial if there is salt, ethanol, or fluorescent dye present. It is corrected using the technique of Daviero et al. (2001).

Finally, the tracer concentrations are obtained from the relationship between image gray scale level and dye concentration. This relationship is linear but the constant of proportionality must be measured. This was done as follows. Immediately after each experiment, the solution in the test tank was stirred to make sure it was homogeneous and images were obtained. Because the source dye concentration is known and the inflow rate is monitored, the average dye concentration in the tank can be calculated and the coefficient of proportionality computed. The coefficient varies with distance from the camera and therefore for each image plane. As shown in Figure 2.5, the coefficient increases with distance from the central plane. This variation is caused by the circular shape of the test tank and the differing path lengths through water of the emitted light to the camera. Laser beams off the central plane pass through longer distances of

Lucite due to the curved cylinder wall, which increases beam attenuation. Individual calibration coefficients for each laser sheet were therefore measured and used in computing the tracer concentrations.

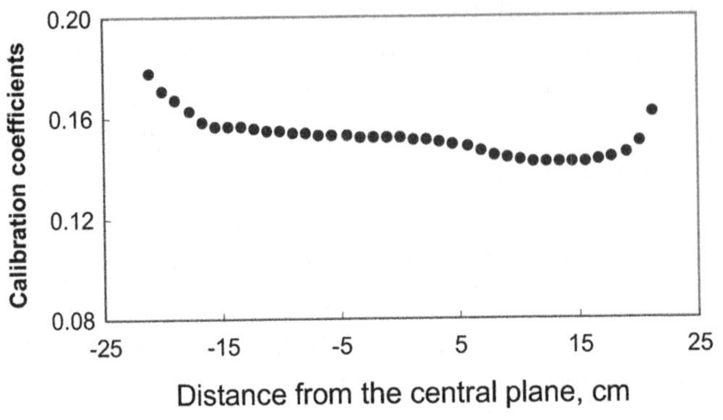

Figure 2.5 Variation of LIF calibration coefficient with distance from vertical tank center plane

Image capture and storage was controlled by the computer program Video Savant™ (IO Industries). This software can capture only raw LIF images, however, so a custom software package, TFLOOK, was developed for calibration, image corrections, animations, and data analyses. TFLOOK is a 32-bit Windows program written in Visual C++ 6.0, for details, see Tian (2002). Finally, two and three-dimensional visualizations of the data were made using TecPlot (TecPlot, Inc.) computer graphics software.

CHAPTER 3.
EXPERIMENTAL RESULTS

INTRODUCTION

The mixing in a tank can be quantified by the coefficient of variation (COV):

$$\text{COV} = \frac{\text{Standard deviation of tracer concentrations}}{\text{Mean tracer concentration}} \tag{3.1}$$

Shortly after the inflow begins, the distribution of tracer in the tank is quite heterogeneous and the COV is large. It decreases as the tank becomes increasingly well-mixed, and approaches zero as the tank becomes fully mixed.

Following the dimensional analyses of Chapter 2, the COV for a cylindrical tank mixed by a jet inflow whose density is the same as the stored water (i.e. $B = 0$) can be expressed as:

$$COV = f\left\{\tau, \frac{H}{D}\right\} \tag{3.2}$$

where $\tau = tM^{1/2} / V^{2/3}$ is dimensionless time. Rossman and Grayman (1999) defined the time for full mixing, t_m (see Eq. 2.8) as the time required for the COV to fall to 0.05 (5%). Based on experiments in cylindrical tanks in which the tracer concentrations were measured with conductivity probes, they estimated the value of the constant in Eq. 2.8 as approximately equal to 10.

Unlike tracer studies with point probe measurements that can only obtain small number of measurements, 3DLIF captures the concentration in the entire tank, allowing the COV to be calculated with far greater accuracy. Typically, the images represent sampling at millions of points in the tank, vastly greater than can be obtained by point-probe techniques. The CCD camera introduces random noise, however, which yields a COV of about 3% in a well-mixed tank. Therefore, we take the criterion for judging the reservoir to be "well-mixed" is that the COV be less than 10%. This yields similar results to those presented by Rossman and Grayman, as shown below.

The dimensionless mixing time, τ_m is a useful measure with which to compare the mixing efficiencies of different tank and inlet geometries. In this section, dimensionless mixing times for each test are presented along with summaries of the test conditions. Further discussions and comparisons between the tank geometries are given in Chapter 4. Some of the many flow images that were obtained are presented in this report; additional images, and two and three dimensional animations of some experiments, are on the accompanying CD.

GC TANK EXPERIMENTS

Introduction

Nine configurations of tank style GC were tested. The experiments were conducted with neutral, positive, and negative buoyancies. The experiments with buoyancy effects corresponded to prototype temperature differences of ±10°F. A summary of the experiments and dimensionless mixing times is given in Table 3.1, for more details see Appendix B, Table B.1.

Table 3.1 Summary of Ground Level Cylindrical (GC) Experiments

Configuration	Tank ID	Model parameters			
		Test Number	Density diff.[1] (σ_t-units)	Flowrate (lpm)	Dimensionless mixing time, τ_m
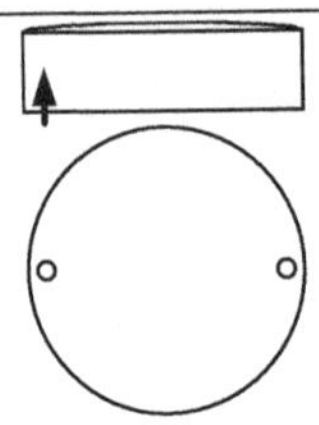	GC01	06	0.0	2.53	11.3
		07	0.0	0.80	9.1
		08	0.0	1.37	9.7
		09	0.0	2.72	11.8
		10	0.0	1.00	8.7
		11	-23.0	1.32	7.5
		12	-23.0	0.50	NM
		13	23.0	1.26	NM
		14	23.0	0.51	NM
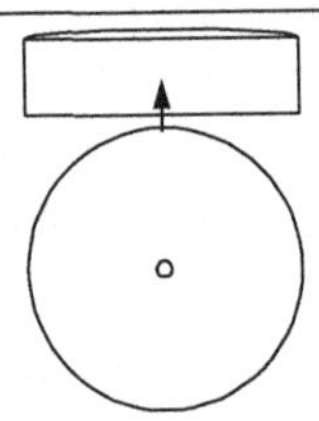	GC02	01	0.0	2.72	11.3
		02	0.0	1.00	20.5
		03	-23.0	1.38	8.0
		04	-23.0	0.52	NM
		05	23.0	1.41	NM
		06	23.0	0.55	NM
		07	0.0	2.53	13.7
		08	0.0	2.16	14.3
		09	0.0	1.80	15.6
		10	0.0	1.43	16.2
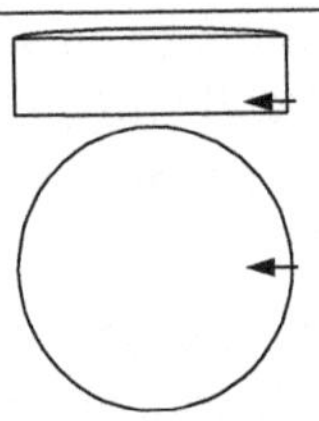	GC03	01	0.0	2.89	11.9
		02	0.0	1.09	10.9
		03	-23.0	1.47	NM
		04	-23.0	0.53	NM
		05	23.0	1.42	9.6
		06	23.0	0.54	NM
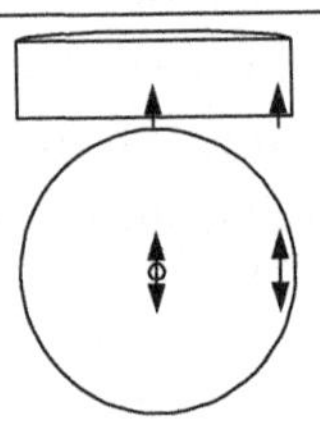	GC04	01	0.0	2.56	11.5
		02	0.0	1.09	11.5
		03	-23.0	1.45	9.2
		04	-23.0	0.53	9.6
		05	23.0	1.43	13.1
		06	23.0	0.52	NM
		07	0.0	2.41	10.8
		08	0.0	1.10	11.1
		09	-23.0	1.43	8.9
		10	-23.0	0.53	NM
		11	23.0	1.43	11.2
		12	23.0	0.52	NM

(continued)

Table 3.1 Summary of Ground Level Cylindrical (GC) Experiments (Continued)

Configuration	Tank ID	Model parameters			
		Test Number	Density diff.[1] (σ_t-units)	Flowrate (lpm)	Dimensionless mixing time, τ_m
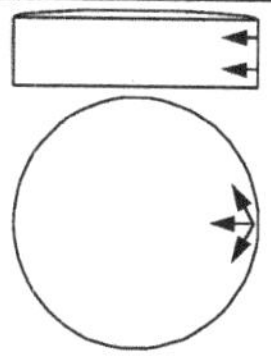	GC05	01	0.0	2.60	9.4
		02	0.0	1.08	8.9
		03	0.0	2.56	9.0
		04	0.0	1.08	9.3
		05	-23.0	1.41	8.4
		06	-23.0	0.54	NM
		07	23.0	1.44	8.6
		08	23.0	0.53	NM
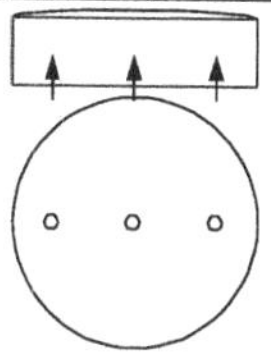	GC06	01	0.0	2.59	7.7
		02	0.0	1.08	8.7
		03	-23.0	1.42	7.3
		04	-23.0	0.54	7.8
		05	23.0	1.39	7.7
		06	23.0	0.53	NM
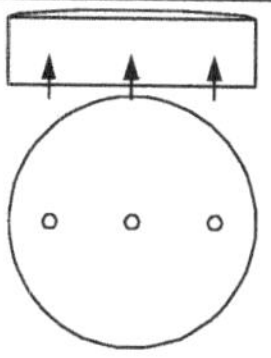	GC07	01	0.0	2.53	8.7
		02	0.0	1.08	8.0
		03	-23.0	1.41	7.6
		04	-23.0	0.55	8.7
		05	23.0	1.42	12.3
		06	23.0	0.53	NM
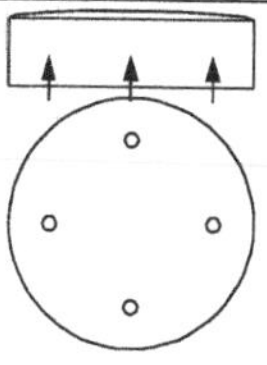	GC08	01	0.0	2.70	-
		02	0.0	1.08	6.1
		03	0.0	2.58	6.3
		04	-23.0	1.39	6.8
		05	-23.0	0.55	8.0
		06	23.0	1.42	7.1
		07	23.0	0.52	NM
		08	-23.0	0.55	7.9
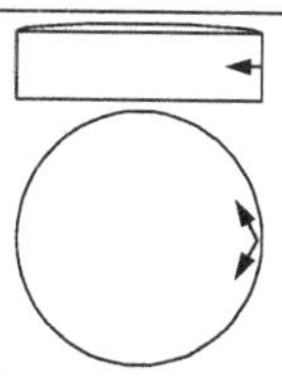	GC09	01	0.0	2.74	8.9
		02	0.0	1.08	8.7
		03	-23.0	1.43	8.3
		04	-23.0	0.54	NM
		05	23.0	1.43	9.6
		06	23.0	0.52	NM
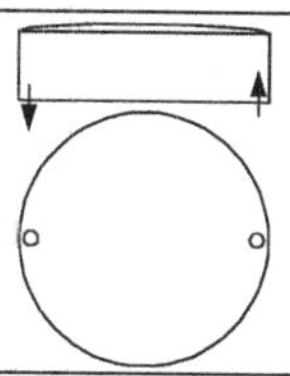	GC01 (With inflow and outflow)	15	0.0	2.00	13.5
		16	0.0	1.07	13.7
		17	0.0	1.06	-
		18	0.0	1.07	13.1
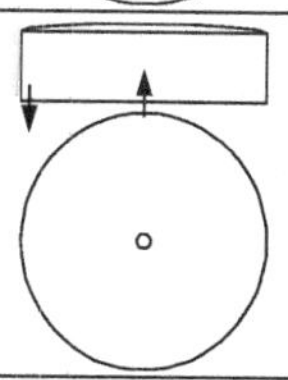	GC02 (With inflow and outflow)	11	0.0	2.00	14.2
		12	0.0	1.07	14.3
		13	0.0	1.06	12.0
		14	0.0	2.14	14.1

[1] A positive density difference means that the inflowing water is warmer than that in the tank, i.e. the inflow is positively buoyant.

NM = Did not become mixed

Experiments with no buoyancy effects

Images of a typical experiment with one nozzle (GC01) are shown in Figures 3.1 and 3.2. Figure 3.1 shows three-dimensional instantaneous visualizations of the tracer concentrations in false color. Figure 3.1a shows three planes out of the 40 obtained, and Figure 3.1b shows how concentration distributions in two orthogonal planes. Figure 3.2 shows an instantaneous tracer concentration distribution on the central plane through the jet. The inflowing jet is the region of high concentration to the left of the image. A recirculation zone can be seen, and three-dimensional effects result from flow around the cylinder walls. Animations of this experiment are on the CD.

The COV and inflow rate are shown in Figure 3.3 as functions of both real time and normalized time. The COV initially increases as the inflow enters the tank, reaches a maximum, and then decreases as the tank becomes uniformly mixed. For this example, the COV becomes equal to 10% after about 85 seconds, corresponding to a dimensionless mixing time of about 10.7.

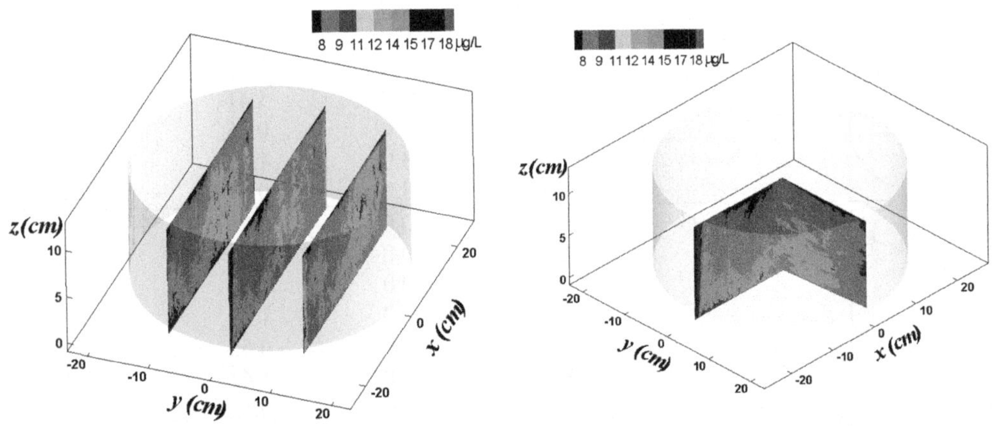

a) Three parallel planes **b) Two orthogonal planes**

Figure 3.1 False-color images of instantaneous tracer concentration (GC01-06)

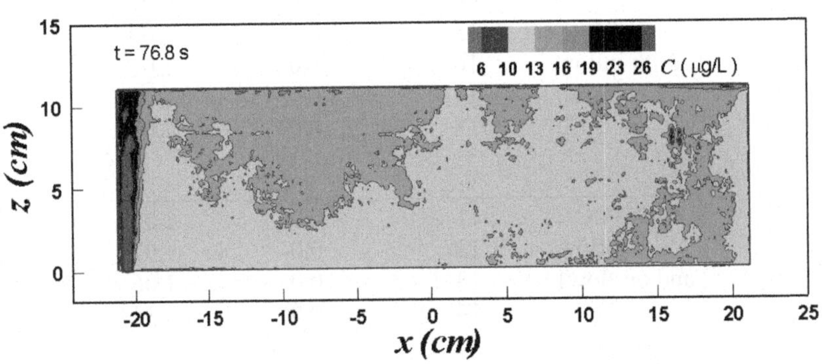

Figure 3.2 False-color image of instantaneous tracer concentration on center-plane (GC01-06)

The COV variation for this and other experiments with tank style GC01 with different inflow rates are plotted versus dimensionless time in Figure 3.4. Although the data are scattered in the early stages, they collapse to a common curve at later times. The dimensionless mixing time, $\tau_m \approx 11$, close to the value of 10.2 suggested by Rossman and Grayman (1999).

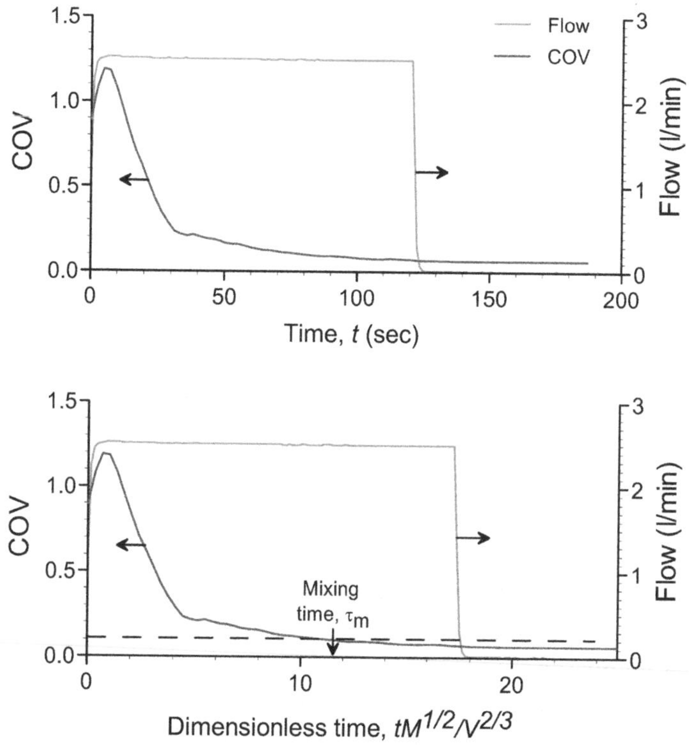

Figure 3.3 Variation of COV and inflow (GC01-06)

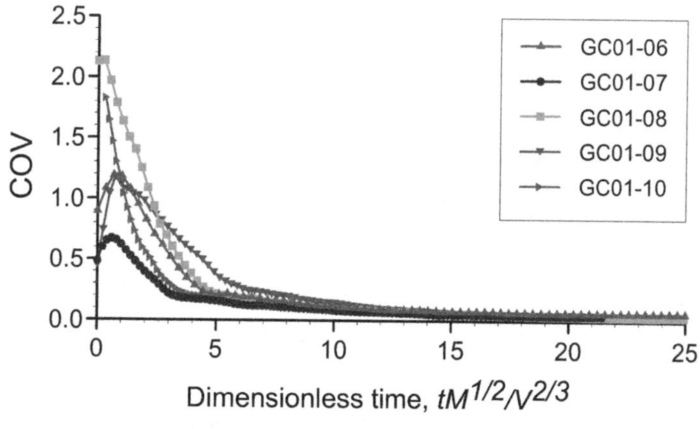

Figure 3.4 Normalized COV measurements for GC01 tests with no density differences

The mixing times for all the GC tank experiments with no buoyancy were computed. They are plotted versus inflow velocity in Figure 3.5. With one exception, the tanks became mixed within the duration of the experiment in all cases. The most rapid mixing occurred with GC08 (four vertical nozzles); the slowest was GC03 (single horizontal nozzle near the floor). The difference between the fastest and slowest mixing times is a factor of about two.

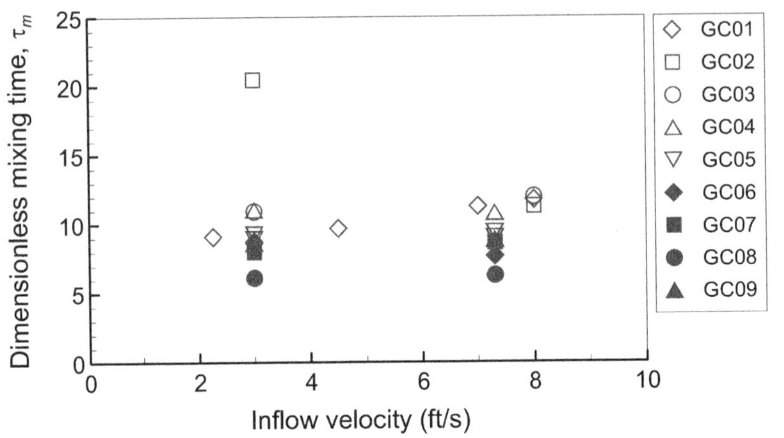

Figure 3.5 Mixing times for GC tanks, no buoyancy

The exception that did not become mixed was test GC02-02. This tank has a single, vertical nozzle in the center of the tank. The reason it did not mix can be seen in Figure 3.6. These images show a donut, or ring-shaped, dead zone that contains low tracer concentrations. This dead zone was quite stable and persistent, as can be seen on the animations on the CD, preventing the tank from becoming mixed. For the same configuration with high momentum (GC02-01), the dead zone was less significant, and the tank did mix. Because of the unexpected nature and importance of this phenomenon, it was studied further; the results are discussed below.

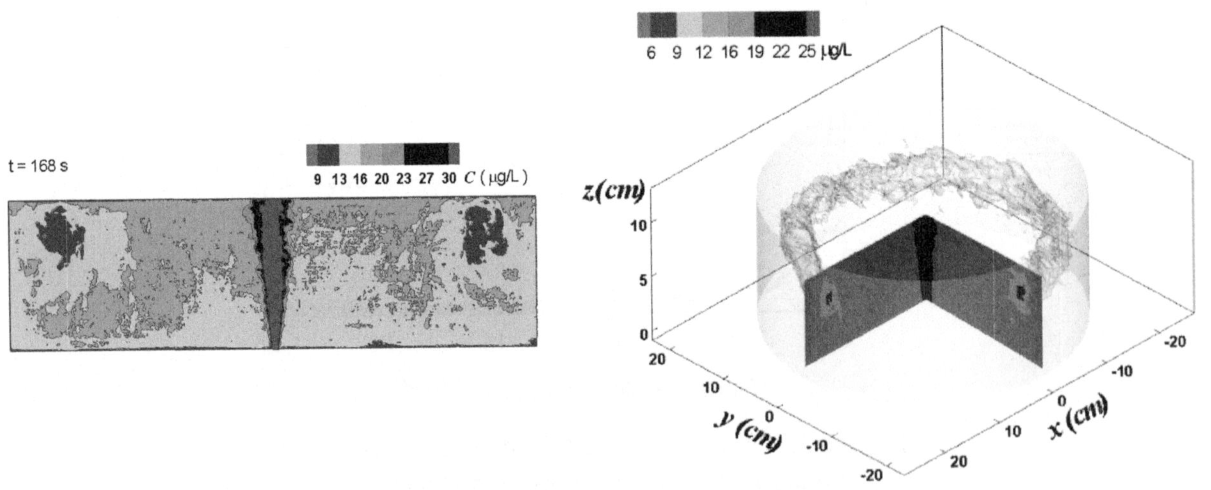

Figure 3.6 Dead zone in experiment GC02-02

Experiments with buoyancy effects

The experiments with buoyancy effects were conducted with density differences corresponding to temperature differences between the inflow and the stored water of about ±10°F.

The dimensionless mixing times are shown in Figure 3.7. For high momentum inflows, the tanks again became mixed with little apparent effect of the density differences. Smaller momentum fluxes, however, could not overcome the effects of the density differences and the tanks did not mix for positively buoyant jets; the tanks became stratified. For negatively buoyant jets, the tank mixed for single-nozzle configurations, but not for multiple nozzles.

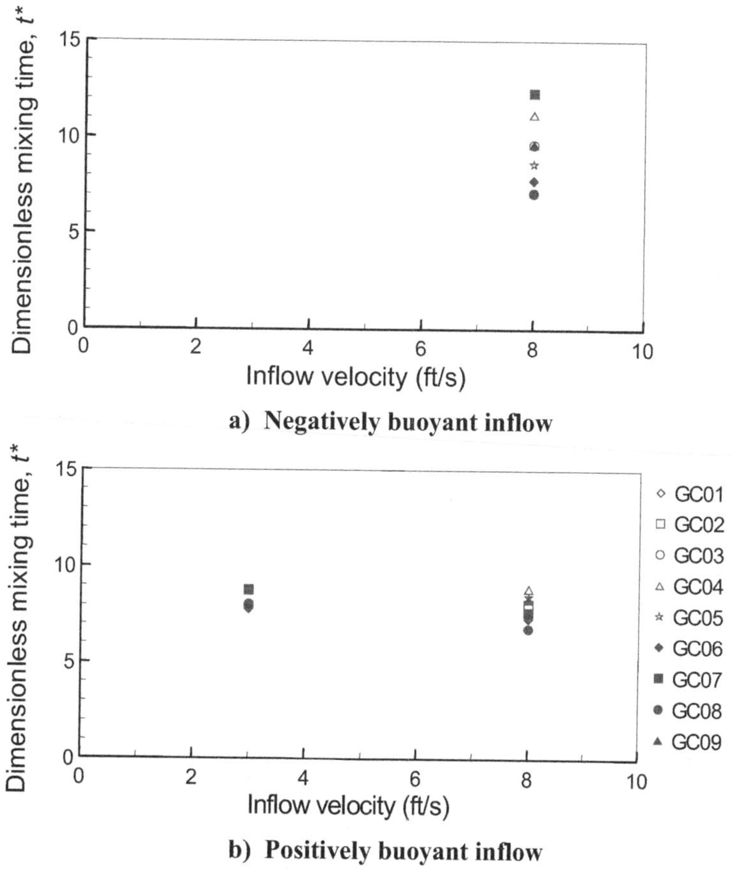

Figure 3.7 Mixing times for GC tanks with density differences

Center-plane images for the experiments that did not mix and became stratified are shown in Figure 3.8. Some of the resulting stratifications were quite strong, for example GC03-04 (single, horizontal nozzle near bottom, with low flowrate and negative buoyancy), and most of the inflow stayed near the bottom.

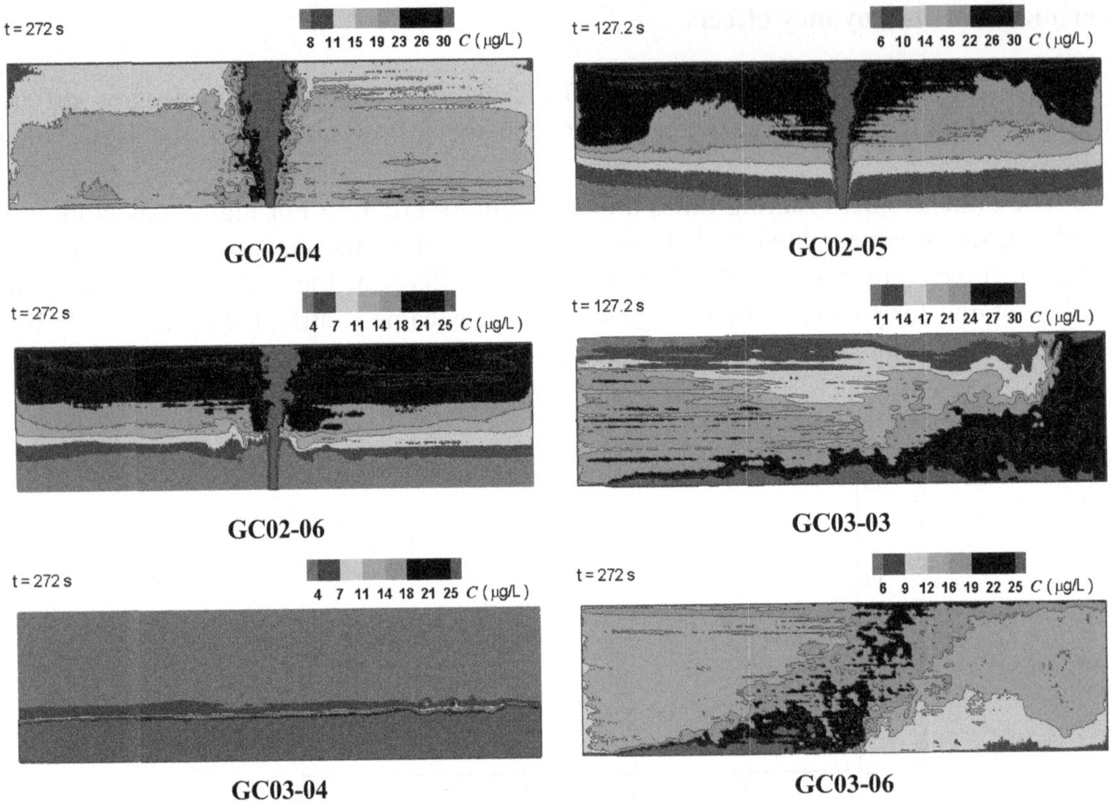

Figure 3.8 Center plane images for GC tank experiments that became stratified

Whether the inflow mixes the tank depends on the value of l_M/H, or $F_j/(H/d)$, where $l_M = M^{3/4}/B^{1/2}$ is a length-scale and F_j is the densimetric Froude number of the jet defined in Chapter 2. The following observations apply for the present experiments $(0.38 < F_j/(H/d) < 1.7)$.

When $F_j/(H/d) < 0.63$, for the tanks with one or two nozzles, mixing did not occur for either positively or negatively buoyant inflows. For tanks with three or four nozzles, the tanks mixed for negatively buoyant inflows but did not mix for positively buoyant inflows. When $F_j/(H/d) > 1.0$, for experiments with more than one nozzle, the tanks mixed for both positively and negatively buoyant inflows. For vertical discharges with one nozzle (GC01 and GC02), the tanks mixed for negatively buoyant inflows but not for positively buoyant inflows. For horizontal discharges with one port (GC03), the tank mixed for positively buoyant inflows but not for negatively buoyant inflows. These observations are discussed further in Chapter 4.

Experiments on the "donut" dead zone

As previously discussed, some of the GC02 experiments (one vertical nozzle at the tank center) with no density differences showed a "donut" shaped dead zone. Because this dead zone precludes complete mixing within the tank, further experiments were performed to investigate its nature. Six experiments were conducted with inflow velocities ranging from 3 to 8 ft/s. We cannot do inflow velocities greater than 8 ft/s in the present experiments.

Images of the experiments are shown in Figure 3.9. The dead zones (the blue/purple regions) are quite apparent as in all experiments except for the one with the highest inflow rate; the dead zone was indeterminate for this experiment.

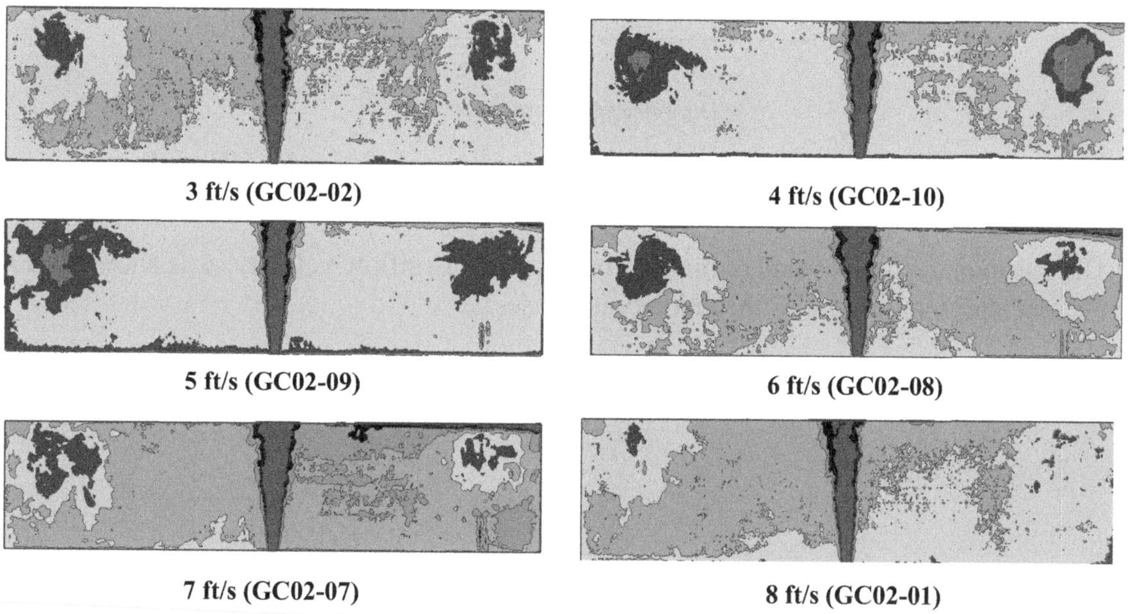

Figure 3.9 Center plane images with central inlet showing dead zones (Blue/purple areas)

Variations of COV with dimensionless times for experiments GC02-01 (8 ft/s) and GC02-02 (3 ft/s) are shown in Figure 3.10. Also shown for comparison are results for tanks with single vertical side inlets (GC01) at the same velocities. For dimensionless times less than about seven, the lower inflow velocities show lower COV values, but the curves approach each other for longer dimensionless times. For the same velocity, the side inlets (GC01) show lower COV values (i.e. better mixing) than the center inlets. Although the mixing time for the center inlet at 8 ft/s was not significantly longer than for the side inlet (according to the criterion COV ≤ 0.1), the center inlet experiments show a longer tail and hence worse mixing for longer times, because of the dead zone. These results indicate that side inlets may be preferable to center inlets.

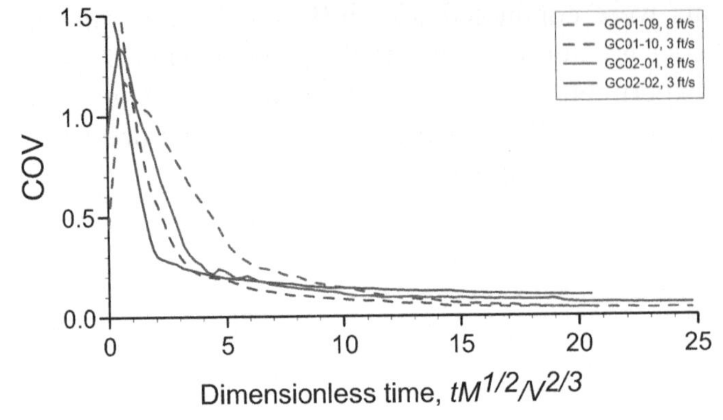

Figure 3.10 Variations of COV for center and side-inlet GC tanks.

The variation of dimensionless mixing time, τ_m versus inflow velocity, u_j is shown in Figure 3.11. τ_m decreases somewhat as the velocity increases.

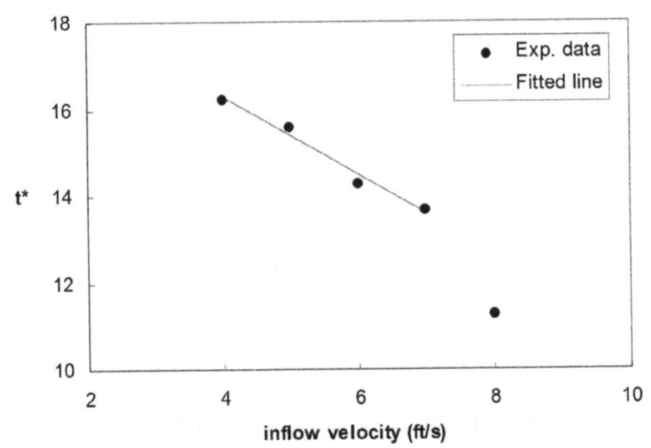

Figure 3.11 Mixing time for tests with dead zones

Simultaneous inflow/outflow experiments

Some experiments were done with simultaneous inflow and outflow at the same flow rates so the volume in the tank remained constant. The experiments were done with tank GC01 in the configuration shown in Figure 3.12. There is a single vertical inlet nozzle near the tank wall and a single outflow of the same diameter near the opposite wall. These experiments were run specifically as test cases for CFD modeling. Four experiments were done at two different flowrates (Table B.3). For test GC01-17 the LIF measurements were made only on the tank center plane. GC01-18 was a repeat of GC01-16.

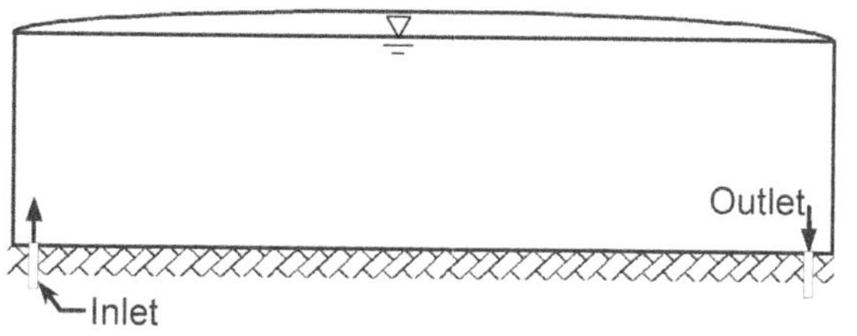

Figure 3.12 Configuration for simultaneous inflow/outflow experiments

Two experiments with and without inflow are shown in Figure 3.13. In both cases, the mixed inflow quickly appeared at the wall opposite to the inlet (near the outlet). With no outflow, it then moved back towards the inlet and was re-entrained, resulting in efficient mixing. With outflow, however, some of the mixed inflow exits through the outlet pipe and recirculation is reduced. This caused the dimensionless mixing times to be longer with outflow than without. With outflow, they were 13.1 to 13.7, compared to 8.7 to 11.8 for similar geometries with no outflow (Figure 3.7 and Table B.3). Mixing times with outflow were similar for the two flowrates tested.

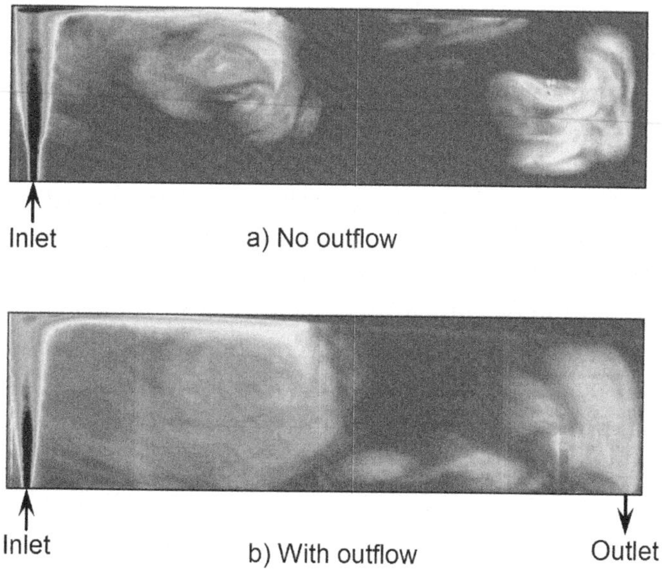

Figure 3.13 Tests with and without simultaneous outflow

Summary

For GC tanks, the mixing time decreases as the number of nozzles is increased. The configuration with four vertical nozzles (GC08) resulted in fastest mixing. The mixing time was not strongly dependent on inlet position or orientation for high momentum flux flows, but did depend somewhat on the inlet position and orientation for low momentum flux. For negatively buoyant jets, vertical nozzles are better; for positively buoyant jets, horizontal nozzles are better. These findings are discussed further in Chapter 4.

ST TANK EXPERIMENTS

Introduction

Experiments were conducted on the Standpipe (ST) model with seven different configurations, as summarized in Table 2.2. Configurations ST01, ST02, and ST03 have single vertical nozzles, and ST07 and ST09 have multiple horizontal nozzles. ST11 has a "draft-tube," that has been proposed as a device to enhance mixing. The nozzle diameters for the multiple nozzle tests, ST07 and ST09, were reduced to maintain the same total port area (and therefore jet velocity and momentum flux) as the single-port tests. Tests ST01 through ST09 were conducted at a nominal scale of 80:1. The prototype tank diameter is then 50 ft and the initial water depth 125 ft. For the single nozzle cases the prototype port diameter was 16 inches.

The combinations of inflow momentum flux and density differences that were tested are summarized in Table 3.2. The inflow was either "low" or "high" momentum, the density difference was either zero, or corresponded to prototype temperature differences of -2.5, -5.0, or -10.0°F. The inflows were always negatively buoyant (i.e. the inflow was more dense than the water in the tank). For ST01 through ST03, two experiments (low and high momentum) were conducted with no density difference. For the experiments with density differences only low momentum inflows were investigated as this is the most critical case for mixing. Two experiments were conducted for each multiport model (ST07 and ST09), both with low momentum. One had no density difference and the other a density difference corresponding to a temperature difference of -10°F. For details of the experiments, see Table B.2.

Tests ST11 and 12 were conducted to investigate the effects of a draft tube on mixing. The model is geometrically similar to an actual draft tube design in order to replicate the claimed low pressure at the inlet that entrains flow into the draft tube. Because the prototype dimensions for the draft-tube tank were slightly different than for the tests above, the model scale was changed to 85.7:1. The prototype dimensions are shown in Figure 3.14. ST12 is identical to ST11 except there was no draft tube. Five tests were conducted for each configuration: high and low momentum with no density difference, and low momentum jets with negatively buoyant inflows corresponding to prototype temperature differences of -2.5, -5.0, or -10.0°F.

Table 3.2 Summary of Standpipe (ST) Experiments

Configuration	Tank ID	Model parameters			
		Test number	Density diff.[1] (σ_t)	Flowrate (lpm)	Dimensionless mixing time, τ_m
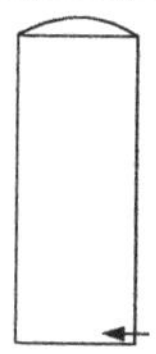	ST01	01	0.0	1.09	18.6
		02	0.0	2.58	18.1
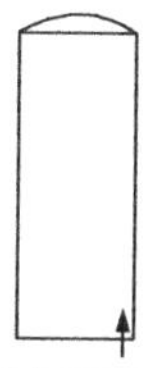	ST02	01	0.0	1.09	15.9
		02	0.0	2.20	14.8
		03	-23.0	0.55	NM
		04	-10.8	0.57	NM
		05	-5.2	0.55	NM
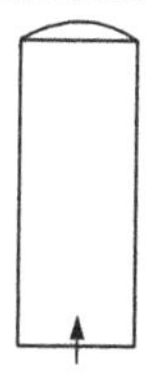	ST03	01	0.0	2.55	15.1
		02	0.0	1.07	15.2
		03	-5.2	0.56	NM
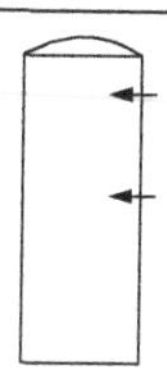	ST07	01	0.0	1.11	13.9
		02	-23.0	0.56	7.3
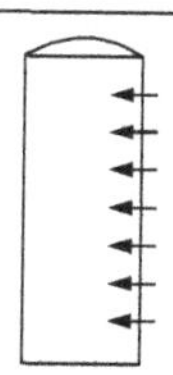	ST09	01	0.0	1.11	13.0
		02	-23.0	0.55	NM
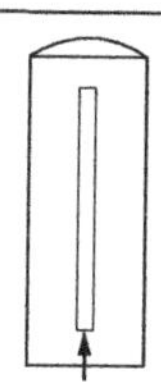	ST11	01	0.0	2.30	NM
		02	0.0	1.07	NM
		03	-23.0	0.55	NM
		04	-10.8	0.55	10.7
		05	-5.2	0.56	12.5
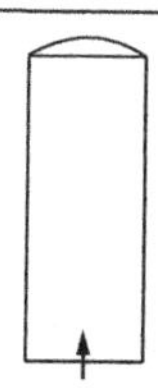	ST12	01	0.0	2.34	15.8
		02	0.0	1.07	15.5
		03	-23.0	0.56	NM
		04	-5.2	0.56	13.4
		05	-10.8	0.56	NM

[1] A positive density difference means that the inflowing water is warmer than that in the tank, i.e. the inflow is positively buoyant.

NM = Did not become mixed

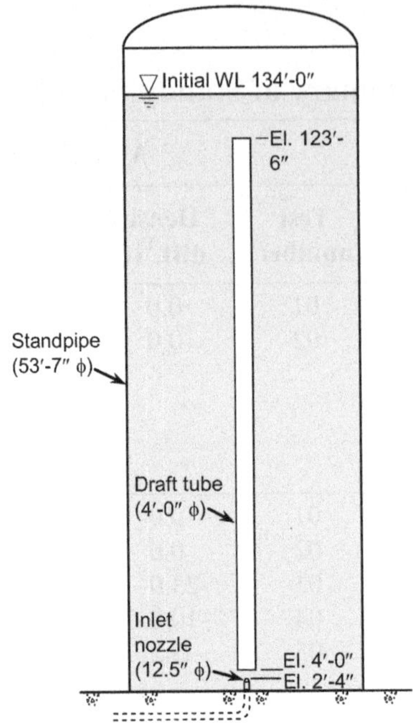

Figure 3.14 Prototype dimensions of standpipe with draft tube (ST11)

Experiments with no buoyancy effects

Dimensionless mixing times for the experiments that became mixed are shown in Figure 3.15. The dimensionless mixing times range from about 13 to 19 and do not show a dependence on inflow velocity. These times are generally longer than those observed for ground level cylindrical tanks (GC) with single nozzles, in other words, ST tanks are more difficult to mix than GC tanks. For a single nozzle, the horizontal orientation (ST01) takes significantly longer to mix than vertical (ST02 and ST03), and the vertical center nozzle (ST03) mixed slightly faster than the vertical side nozzle (ST02). With the draft tube, the tank did not become well-mixed for either high or low momentum inflows; this is discussed further below.

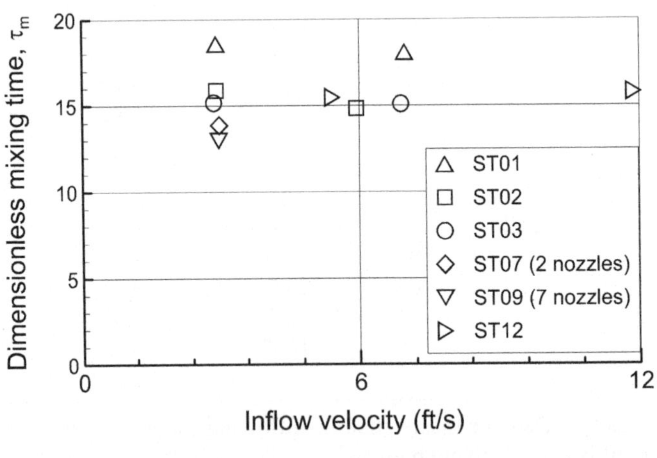

Figure 3.15 Mixing times for ST tanks – no buoyancy

Increasing the number of nozzles resulted in somewhat more rapid mixing. The mixing time decreased by about 8 to 15% compared to the vertical single nozzles; seven-nozzles (ST09) were slightly better than two (ST07).

The reason for the poor mixing of the horizontal nozzle (ST01) can be seen in the LIF image, Figure 3.16. The flow separates into two regions, with higher concentrations in the lower region, and lower concentrations in the upper region. Mixing between these two regions is slow, resulting in longer mixing times.

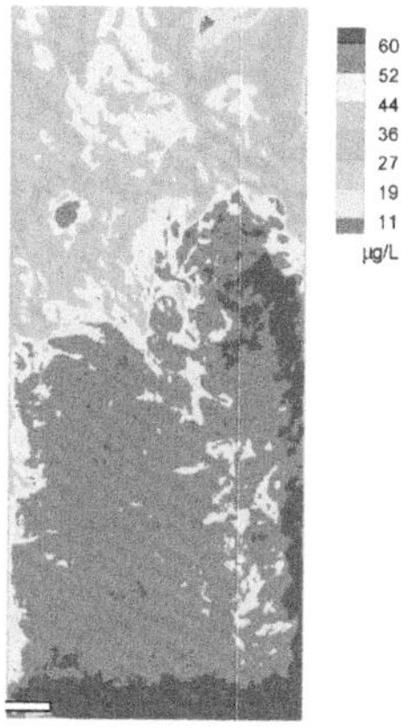

Figure 3.16 Horizontal nozzle (ST01-01)

With a draft tube, the tank did not mix for either the high or low momentum inflows. The reason can be seen in Figure 3.17 which shows flows at the same times with and without a draft tube. (The top of the tube is outside the field of view). Entrainment into the jet is reduced by the tube and can only occur over the exposed length of the jet before it enters the tube. The mixed inflow exits from the tube top, and then spreads out and slowly sinks and mixes with the ambient water. The concentration distribution is clearly more spatially heterogeneous than without the tube, i.e. the tank is less well-mixed.

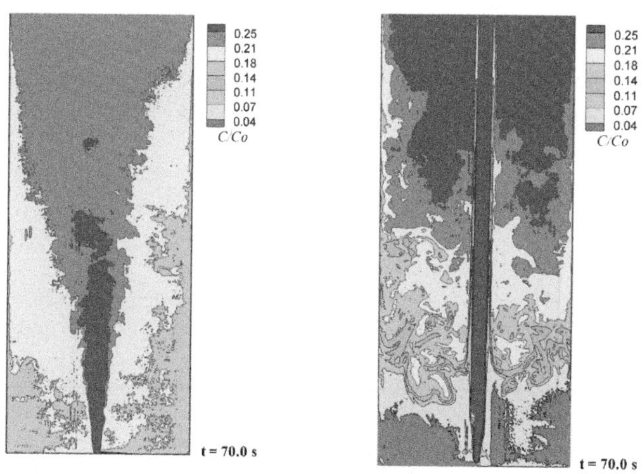

Without draft tube (ST11-01) **With draft tube (ST12-01)**

Figure 3.17 Effect of draft tube on mixing - no buoyancy

This behavior can be quantified by comparing the COVs plotted in Figure 3.18. The COV for the experiments without the draft tube decrease rapidly, indicating efficient mixing, and becomes well-mixed at a dimensionless time of about 15. In contrast, the COV with the draft tube remain high, indicating large spatial variability, and do not fall below 10%, the assumed criterion for the tank to be "well-mixed."

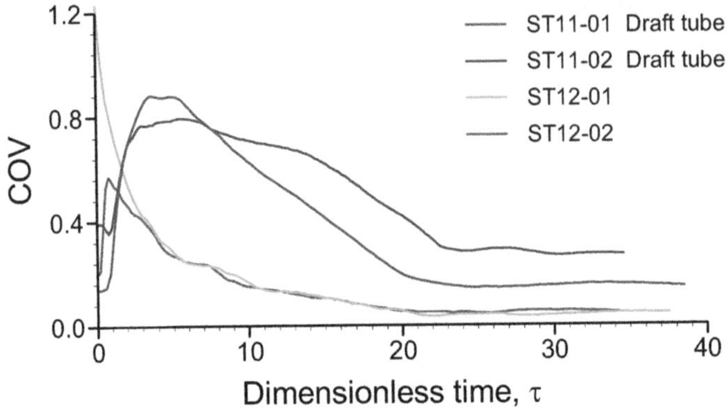

Figure 3.18 Mixing with and without a draft tube

Experiments with buoyancy effects

The ST experiments with buoyancy effects were conducted with density differences corresponding to prototype temperature differences of -2.5, -5.0, and -10.0°F. The inflows were negatively buoyant, i.e. more dense than the water in the tank.

The dimensionless mixing times for the tanks that became mixed are shown in Figure 3.19. None of the single-nozzle configurations (ST01 through ST03) resulted in uniform mixing. This is because the vertical negatively buoyant jets did not reach the water surface, and mixing only occurred below the maximum rise height.

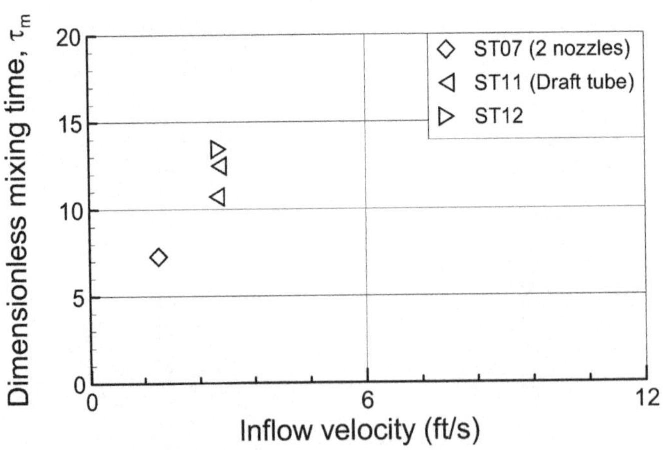

Figure 3.19 Mixing times for ST tanks with negatively buoyant inflows

The maximum rise height, z_m of a vertical negatively buoyant jet is proportional to the product of the nozzle diameter, d and the jet densimetric Froude number, F_j (Roberts and Toms, 1987):

$$z_m = CdF_j \tag{3.3}$$

where C is an experimental constant ≈ 2.1. The maximum rise height predicted by Eq. 3.3 for experiments ST02 and ST03 is between 36 ft and 76 ft in the prototype, much lower than the initial water depth of 125 ft. The water depth for experiment ST12-04 is 134 ft and z_m is predicted to be 140 ft. The tank did become mixed in this case. These results suggest that an approximate criterion for the tank to mix is $F_j/(H/d) > 0.5$. The rise height for ST01 is zero because the jet is discharged horizontally; therefore, the tank would not be expected to mix.

The draft tube did result in the tank becoming mixed for some of the experiments with density differences. The reason can be seen in Figure 3.20, which shows images for a temperature difference of -5°F with and without the draft tube. The flow up the draft tube can be clearly seen as the red column. For this density difference (and smaller), the inflow exited the top of the tube with considerable momentum. It then spread out and ascended (because the mixture is heavier than the ambient water) resulting in rapid mixing. When the density difference was large (10°F), however, the inflow only just reached the top of the tube. It had little residual momentum and sank to the tank bottom with little mixing. The top of the draft tube is above the field of view of this image but some of the mixed inflow can be seen falling down outside the tube. In contrast the tank without the tube shows that the dense jet (the red area) does not reach the water surface. It falls back on itself and fills the tank from the bottom, resulting in strong stratification that resists further mixing. The role of the draft tube in this case is to provide a path that does not otherwise exist for the jet to the top of the tank. See the CD for animations of these experiments.

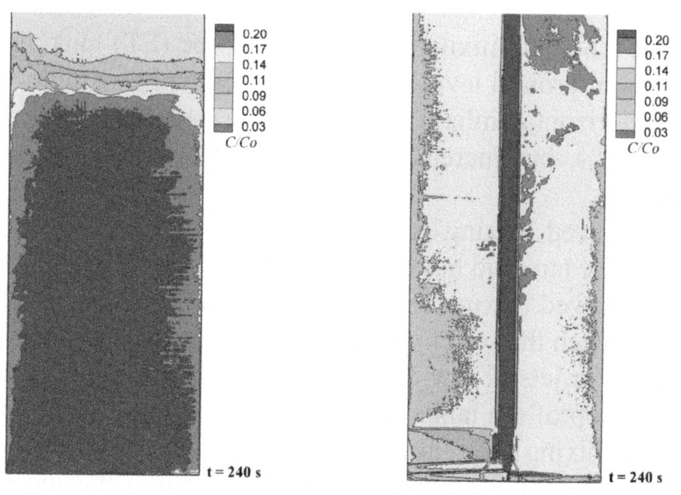

Without draft tube (ST12-05) **With draft tube (ST11-04)**

Figure 3.20 Effect of draft tube with negatively buoyant inflows

Multiport nozzles again resulted in more rapid mixing. As previously discussed, with no density difference, seven-nozzles were slightly better than two. With a density difference, however, two nozzles (ST07) were better, and the seven-nozzle tank (ST09) did not become well mixed. The reasons can be seen in Figure 3.21. The top nozzle in ST09 is higher, resulting in mixing up to the water surface, and a long entrainment path as the jet falls to the bottom. The concentration in ST07 is fairly uniform but in ST09 it is layered with different concentration levels. These results are discussed further in Chapter 4.

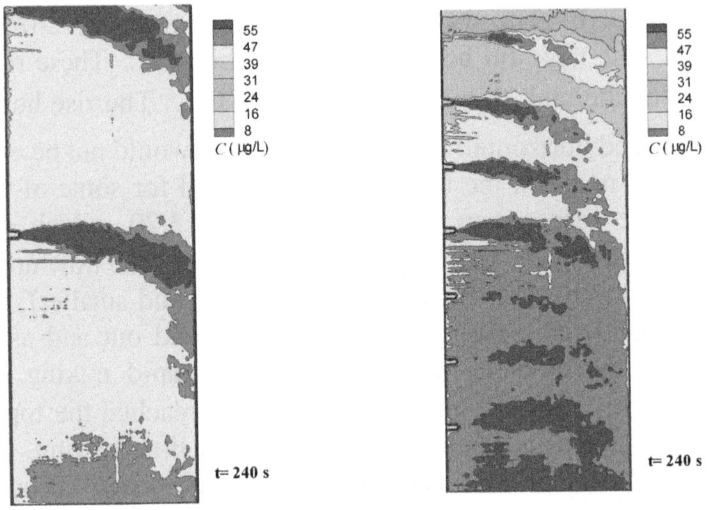

ST07 **ST09**

Figure 3.21 Two- and seven-nozzle tanks at $t = 240$ s

Summary

Without the draft tube, the mixing time for standpipe (ST) tanks is generally longer than for similar configurations of ground level cylindrical (GC) tanks. Single vertical inlets are more efficient than a single horizontal inlet. Increasing the number of nozzles to two decreased mixing times by about 10%, but increasing the number of nozzles beyond two only slightly reduced the mixing times.

The draft tube reduced mixing with no density difference because the tube inhibited entrainment into the jet. The tank did not become well-mixed with the draft tube installed.

The draft tube improved mixing with negatively buoyant inflows, however. Without the draft tube, the jet did not reach the water surface, resulting in strong stratification in the tank that inhibited further mixing. For jets of low or medium density difference, the draft tube provided a pathway for the jet to the top of the tank. The residual momentum and excess density of the jet then caused it to ascend, mixing the tank. For larger density difference, however, the residual momentum of the jet was too small to mix the tank. Varying the draft tube geometry might improve its mixing capabilities.

With negatively buoyant inflows two nozzles at mid depth and near the water surface mixed more rapidly than did one nozzle.

GR TANK EXPERIMENTS

Introduction

Experiments were conducted on ground level rectangular (GR) model tanks with single and multiple nozzles in the configurations summarized in Table 2.2. The tests were conducted at a nominal scale of 81.6:1. The tanks had 1, 3, or 4 nozzles. For the single nozzle configurations the prototype port diameter was 16 inches. The model sidewall lengths are 15.4″ (i.e. the tanks are square) and the initial water depth was 4.4″. At a scale of 81.6:1 this corresponds to prototype tank sidewall lengths equal to 105 ft and an initial water depth of 30 ft.

The combinations of inflow momentum flux and density difference between the inflow and the tank water that were tested are summarized in Table 3.3. The experiments with density differences corresponded to prototype temperature differences of -2.5, -5.0, or -10.0°F; the inflows were always negatively buoyant (i.e. the inflow was more dense than the water in the tank). Only low momentum inflows were tested with density differences as this is the most critical case. For details of the experiments, see Appendix B, Table B.3.

Experiments with no buoyancy effects

The results for tests with no density differences are shown in Figure 3.22 as the dimensionless mixing time plotted versus the number of nozzles. The tanks became well mixed for all cases. The mixing time generally decreases with increasing number of nozzles. For a single nozzle, the horizontal orientation is most efficient. The vertical nozzle (GR04) results in considerably longer mixing time. This is probably because the jet interacts with the water surface and creates surface fluctuations that apparently impede circulation and mixing, particularly at high influx momentum. This effect can be seen in Figure 3.23 where, for the same dimensionless time, the mixed inflow has not yet reached the far side. Similar results were observed for the ground level cylindrical (GC) tank except that in that case, the mixing time with high momentum was only about 20% higher. Otherwise, the mixing times with rectangular tanks were generally comparable to those for round tanks.

The horizontal orientation was also better with multiple nozzles. This is because the tanks are long and relatively shallow, so the horizontal jet affords greater opportunity for entrainment before impacting a boundary than do the vertical or 45° nozzles. The best configuration was GR05, which has four ports arranged in a cluster discharging horizontally near the center of the tank at mid depth.

Table 3.3 Summary of Ground Level Rectangular (GR) Experiments

Configuration	Tank ID	Model parameters			
		Test number	Density diff.[1] (σ_t-units)	Flowrate (lpm)	Dimensionless mixing time, τ_m
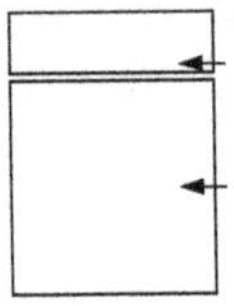	GR01	01	0.0	1.08	9.2
		02	0.0	2.79	10.6
		03	-5.2	0.53	NM
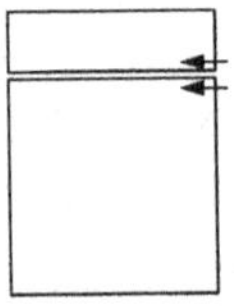	GR02	01	0.0	1.06	9.5
		02	0.0	2.80	9.9
		03	-5.2	0.53	NM
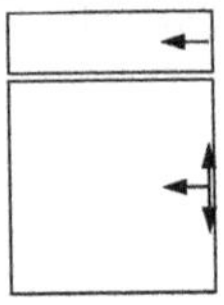	GR03	01	0.0	1.09	7.4
		02	0.0	2.74	7.4
		03	-5.2	0.54	NM
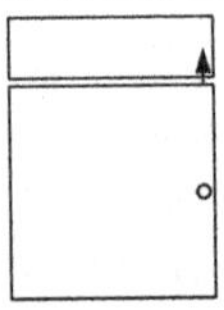	GR04	01	0.0	1.08	8.7
		02	0.0	2.80	15.5
		03	-5.2	0.53	7.8
		04	0.0	2.80	14.7
		05	0.0	2.75	15.4
		06	-10.8	0.55	7.7
		07	-23.0	0.53	NM
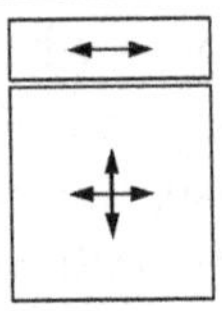	GR05	01	0.0	1.08	6.5
		02	0.0	2.67	5.8
		03	-5.2	0.53	NM
		04	0.0	2.68	5.6
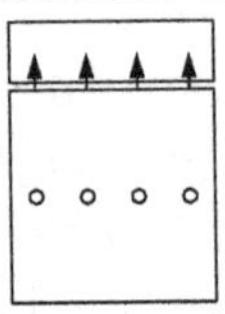	GR06	01	0.0	1.08	10.3
		02	0.0	2.69	9.5
		03	-23.0	0.54	7.5
		04	0.0	1.08	10.1
		05	0.0	2.60	9.4
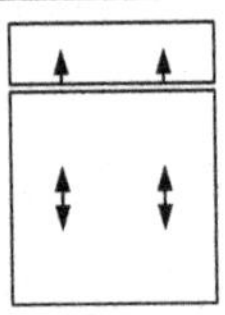	GR07	01	0.0	1.09	7.8
		02	0.0	2.74	7.9
		03	-5.2	0.53	6.3
		04	-10.8	0.54	6.8
		05	-23.0	0.55	NM
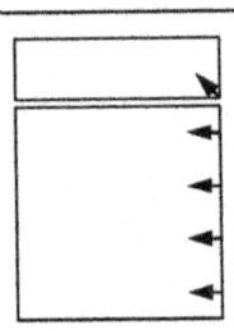	GR08	01	0.0	1.08	9.7
		02	0.0	2.70	9.3
		03	-23.0	0.52	NM
		04	-10.8	0.55	7.3

[1] A positive density difference means that the inflowing is warmer than that in the tank, i.e. the inflow is positively buoyant.

NM = Did not become mixed

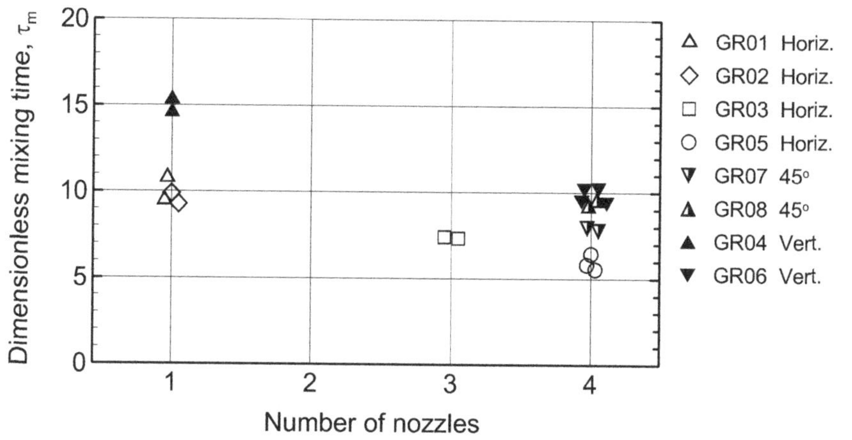

Figure 3.22 Dimensionless mixing times for GR tanks with no buoyancy

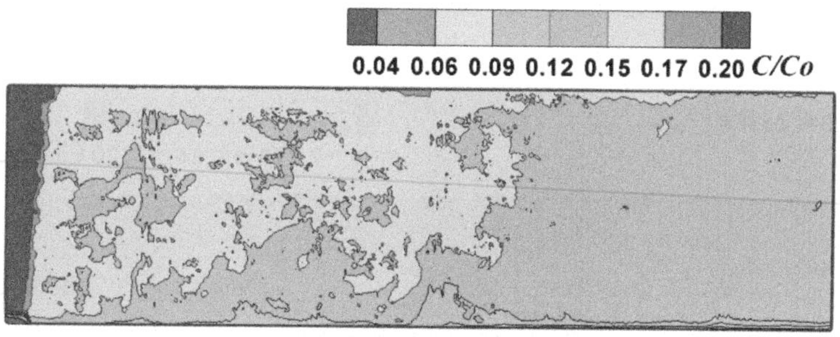

Low momentum (GR04-1)

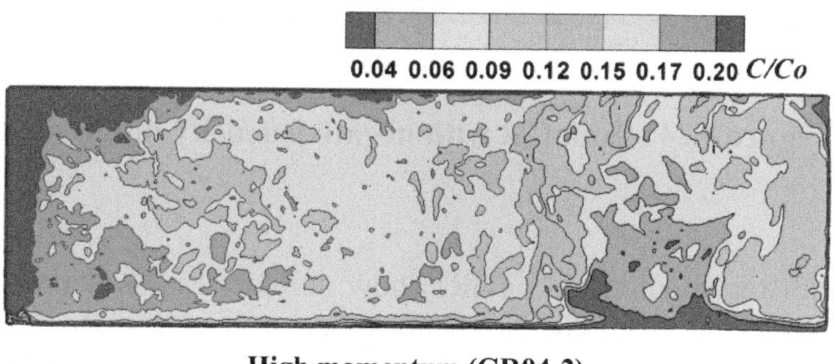

High momentum (GR04-2)

Figure 3.23 Low and high momentum vertical inflows with one nozzle into GR tank (τ = 10.4)

Experiments with buoyancy effects

The experiments with buoyancy effects were conducted with density differences corresponding to prototype temperature differences of about -2.5, -5.0, and -10.0°F.

The horizontal discharge was least efficient at mixing. As shown in Figure 3.24, tanks GR01 and GR02 became strongly stratified and the dense inflow stayed near the tank bottom. Because mixing was not observed for the smallest density differences, larger density differences were not tested. For a vertical nozzle (GR04), the tank became well mixed for small density differences (-2.5 and -5.0°F) but not for larger (-10.0°F). Only tank GR06 (four vertical nozzles) resulted in uniform mixing for the highest density difference.

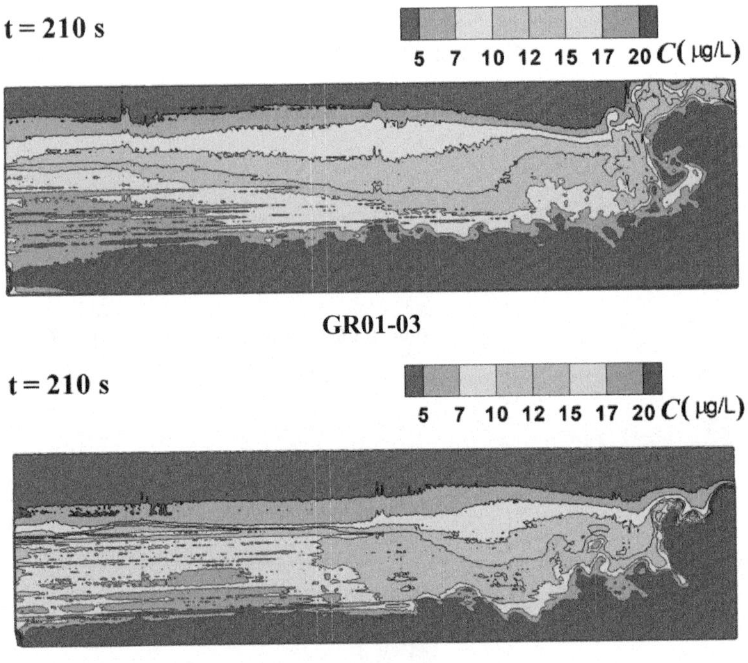

Figure 3.24 Tank GR with negative density differences

Summary

The tests with rectangular tanks (GR) show similar results to those obtained with the cylindrical (GC) tanks. Horizontal nozzles result in best mixing with no density difference, and vertical orientations are best with negatively buoyant inflows. Increasing the number of nozzles results in more rapid mixing.

Effect of depth to diameter ratio for cylindrical tanks

Many water storage tanks are cylindrical. For a given inlet configuration and with no density differences, the dimensionless mixing time is given by Eq. 2.7:

$$\tau_m = \frac{t_m M^{1/2}}{V^{2/3}} = f\left(\frac{H}{D}\right) \tag{3.4}$$

The effect of *H/D* is usually neglected, so the dimensionless mixing time is constant. Rossman and Grayman (1999) performed experiments for cylindrical tanks with single horizontal and vertical nozzles over the range $0.11 < H/D < 0.49$ and obtained:

$$\tau_m = \frac{t_m M^{1/2}}{V^{2/3}} = 10.2 \tag{3.5}$$

Their data had considerable scatter, however, with values ranging from 6.4 to 17.4.

The GC and ST experiments with cylindrical tanks that were previously discussed had constant *H/D* equal to 0.25 and 2.5, respectively. Additional experiments were performed using these tanks with different water levels to systematically investigate the effect of *H/D*. The inflow was discharged vertically from a single nozzle at the bottom near the wall (GC01 and ST02 configurations, see Table 2.2). Only cases with no density difference were tested. For the GC model, experiments were conducted at $H/D = 0.25, 0.50, 0.75, 1.0$ (GC01-06 to GC01-10 and GC01-19 to GC01-22) and for the ST model, experiments were conducted at $H/D = 0.5, 1.0, 1.5, 2.0, 2.5$ (ST02-01 to ST02-02 and ST02-06 to ST02-13). The results are included in Tables B.1 and B.2.

The results are plotted in Figure 3.25. Also shown for comparison is the result of Rossman and Grayman (1999). It can be seen that the dimensionless mixing time becomes longer as *H/D* increases above about 1, i.e. as the tank becomes more tall and slender. For equal volumes and inflow conditions, the mixing time for a tank with $H/D = 2.5$ is about 50% longer than for one with $H/D < 1$. The results can be approximated by:

$$\begin{aligned} &\tau_m \approx 10 \quad \text{for } H/D \le 1 \\ &\tau_m \approx 10 + 3.5\left(\frac{H}{D} - 1\right) \quad \text{for } H/D > 1 \end{aligned} \tag{3.6}$$

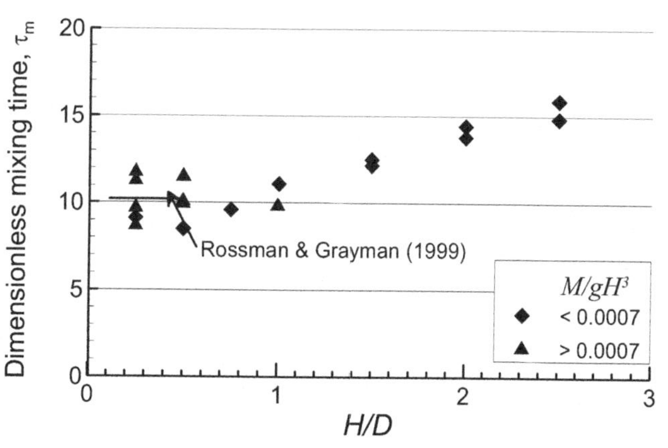

Figure 3.25 Effect of depth-to-diameter ratio on mixing in cylindrical tanks

The results show more scatter at small values of H/D. For these cases, the high momentum inflow caused surface displacements and waves. This is a gravitational phenomenon that is not included in the dimensional analysis leading to Eq. 2.7. An additional parameter in this case is M/gH^3, which is a measure of the extent to which the jet still has sufficient velocity at the water surface to cause surface waves. Figure 3.25 suggests that free surface effects are not important when $M/gH^3 < 0.0007$. Higher values lead to free surface effects that increase the data scatter.

CHAPTER 4.
DISCUSSION OF EXPERIMENTAL RESULTS

INTRODUCTION

The experiments have generated considerable data covering wide ranges of storage tank configurations and flow characteristics. They show the water tank flows to be dominated by large-scale, coherent unsteady vortices, which account for a significant part of the transport and mixing. These flows will be a challenge even for advanced CFD codes. Recent work has shown that eddy-viscosity models fail in flows dominated by coherent structures and accurate predictions of such flows require the use of unsteady statistical turbulence models, based on hybrid unsteady Reynolds averaged equations (URANS) and large eddy simulations (LES) formulations (Paik et al, 2004). Unsteady URANS/LES models, which have shown great promise in complex engineering flows (Paik et al. 2004, 2005), should be extended and applied to water tank flows. Further discussion of the experimental results is presented in this section, in particular of buoyancy effects and to compare tank and inlet geometries.

EXPERIMENTS WITH NO BUOYANCY EFFECTS

In order to compare the mixing efficiencies of various inlet configurations with no buoyancy effects, it is convenient to plot the dimensionless mixing times versus the number of nozzles. The graphs in this section show vertical discharges by solid symbols, horizontal by open symbols, and other orientations by half-filled symbols. For the actual nozzle orientations, see Table 2.2.

The experiments with ground level cylindrical (GC) tanks are summarized in Tables 3.1 and B-1. The following discussion is for a constant ratio of water depth to diameter, *H/D* = 0.25. Other *H/D* values were discussed in Chapter 3. The mixing times are shown in Figure 4.1.

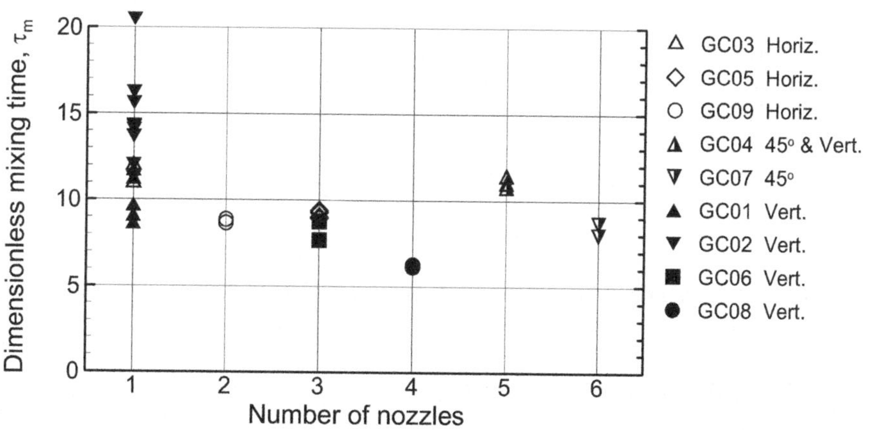

Figure 4.1 Mixing times for GC tanks with no buoyancy effects

All of the nozzle configurations resulted in uniform mixing (as defined here). There were significant differences in the mixing times, however, among the different configurations. The most efficient mixing (lowest dimensionless mixing time) is achieved by the tank GC08, which has four vertical nozzles. The dimensionless mixing time, $\tau_m \cong 6.2$ for this case. Increasing the number of vertical ports would presumably further decrease the mixing time, but this was not tested. The configurations with five and six ports (GC04 and GC07) had ports oriented at 45° to the horizontal. These were less efficient. Decreasing the number of vertical ports to three (GC06) increased the mixing time to about 8.5, which was slightly better than three horizontal nozzles (GC05). This was essentially the same as two nozzles at mid-depth (GC09). For a single nozzle, the best configuration was vertical near a wall (GC01), which had an average mixing time of 10.1. A horizontal single nozzle near the floor (GC03) resulted in a slightly longer mixing time of 11.4. The vertical nozzle at the center had much longer mixing times. This is an anomalous case, however, due to the formation of the donut-shaped dead zone that was discussed in Chapter 3.

For ground level circular tanks, vertical nozzle orientations resulted in best mixing. Mixing improved as the number of nozzles was increased. Compared to a single nozzle, three nozzles reduced the mixing time by about 25%, and four nozzles reduced it by almost 50%. If a single nozzle is used, it may be preferable to place it near the sidewall, as this avoids the dead zone that was observed in some experiments. Horizontal orientations resulted in slightly longer mixing times, but 45° orientations were significantly less efficient, even with more than four nozzles.

The experiments with standpipe (ST) tanks tested are summarized in Tables 3.2 and B-2. The following discussion is for a constant ratio of water depth to diameter, $H/D = 2.5$. Other H/D values were discussed in Chapter 3. The mixing times are shown in Figure 4.2.

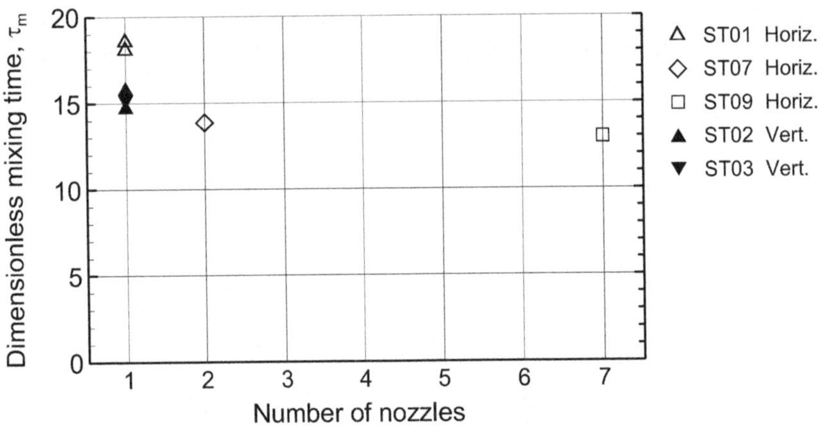

Figure 4.2 Mixing times for ST tanks with no buoyancy effects

Single nozzle configurations were tested in three configurations, all discharging near the floor: Horizontal (ST01), vertical near the wall (ST02), and vertical at the tank center (ST03). The two vertical configurations mixed faster than the horizontal. There was no significant difference between the two vertical nozzle placements, but the horizontal nozzle resulted in about 17% longer mixing time.

Two multiple-nozzle configurations were tested, both with horizontal discharges: Two nozzles at mid depth and near the water surface (ST09), and seven nozzles distributed over the water depth (ST11). The two nozzle configuration reduced the mixing time by about 10% compared to the single nozzle; seven ports only reduced this by an additional 6%. There is therefore little to be gained by increasing the number of nozzles for stand pipe tanks.

The experiments with ground level rectangular (GR) tanks are summarized in Tables 3.3 and B.3. All tests were conducted with square tanks with a ratio of water depth to sidewall length $H/D = 0.29$. The dimensionless mixing times are shown in Figure 4.3.

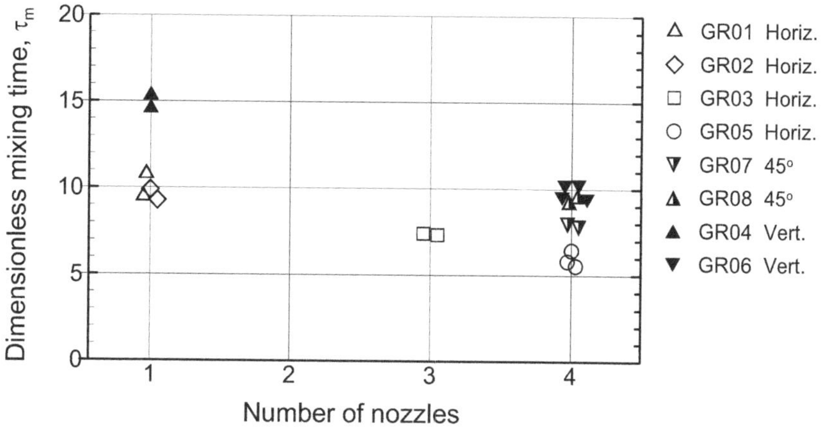

Figure 4.3 Mixing times for GR tanks with no buoyancy effects

Three single nozzle configurations were tested: Horizontal near the floor at the center of a wall (GR01), and near a corner (GR02), and vertical at the center of a wall (GR04). The horizontal nozzles resulted in most rapid mixing, with little difference between nozzle location at the wall center or in a corner. The vertical nozzle was significantly (50%) slower mixing. This is presumably due to the longer entrainment path length for the horizontal nozzles.

Five multiple-nozzle configurations were tested. GR03 had three nozzles clustered together at the center of a wall, mid water depth at right angles to each other. This reduced the mixing time by about 25% compared to the horizontal single nozzles. Four different configurations of four-nozzle arrangements were tested. Again, best results were obtained with horizontal nozzles, GR05, with the nozzles perpendicular to each other at mid depth in the tank center. This configuration reduced the mixing time by 40% compared to the single nozzles. Those with vertical or nozzles at 45° (GR06, GR07, and GR08) compared to the three nozzle configuration.

EXPERIMENTS WITH BUOYANCY EFFECTS

Negative buoyancy

The main variables that determine mixing in a tank where the inflow has a different density than the water in the tank are the momentum flux, M, the buoyancy flux, B, the water depth, H, the tank diameter (or wall length for a square tank), D, the number of nozzles, n, and the inlet geometry. Thus, a more general statement of Eq. 2.7 for the mixing time is:

$$\frac{t_m M^{1/2}}{V^{2/3}} = f\left\{\frac{H}{D}, \frac{M^{1/2}}{B^{1/3}H^{2/3}}, n, \text{Inlet geometry}\right\} \tag{4.1}$$

(Note that $l_M/H = \left(M^{1/2}/B^{1/3}H^{2/3}\right)^{3/2}$ so either group can be used).

Density differences may prevent the tank from becoming mixed. This depends on the values of the independent parameters in Eq. 4.1, particularly the momentum flux parameter. The momentum flux and orientation of inclined nozzles can be approximately combined into the vertical component of momentum, $M\sin\theta$ where θ is the angle of the nozzle to the horizontal. Eq. 4.1 then becomes, for particular nozzle locations:

$$\frac{t_m M^{1/2}}{H^2} = f\left\{\frac{(M\sin\theta)^{1/2}}{B^{1/3}H^{2/3}}, n\right\} \tag{4.2}$$

The results for all tanks with bottom inlets are shown in Figure 4.4. The experiments that became mixed are shown with solid symbols, and those that did not with open symbols.

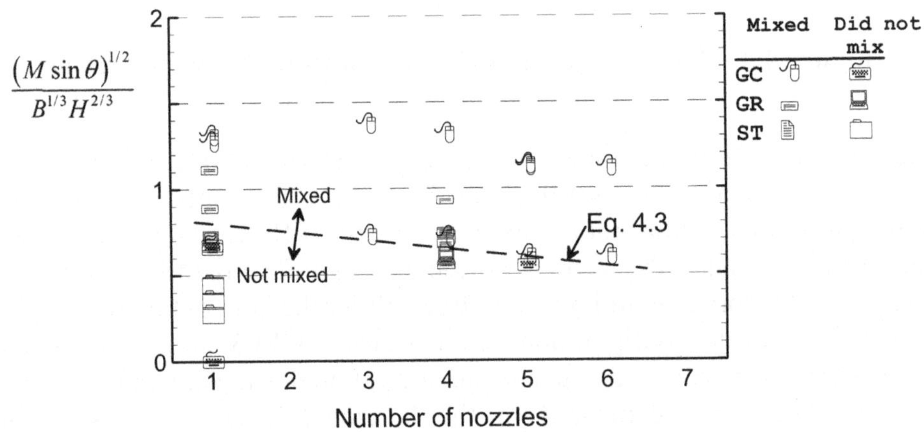

Figure 4.4 Mixing of negatively buoyant inflows for all tanks with bottom inlets

The main parameter that determines whether mixing occurs is the momentum flux parameter of Eq. 4.2 with a weak dependence on the number of nozzles. This is because the main mechanism for mixing is the total momentum flux (although when mixing does occur, it is more rapid with more nozzles, as previously discussed). Figure 4.4 indicates that a simple criterion for the tank to become mixed with bottom inlets is:

$$\frac{(M\sin\theta)^{1/2}}{B^{1/3}H^{2/3}} > 0.85 - 0.05n \tag{4.3}$$

None of the experiments with single, horizontal nozzles became mixed. For horizontal inlets, the vertical component of momentum is always zero so the parameter of Eq. 4.2 is not useful. The total momentum flux parameter of Eq. 4.1 is more appropriate, and the results for tanks with horizontal nozzles are plotted using this parameter in Figure 4.5. This figure includes all experiments with elevated nozzles, in addition to the bottom nozzles.

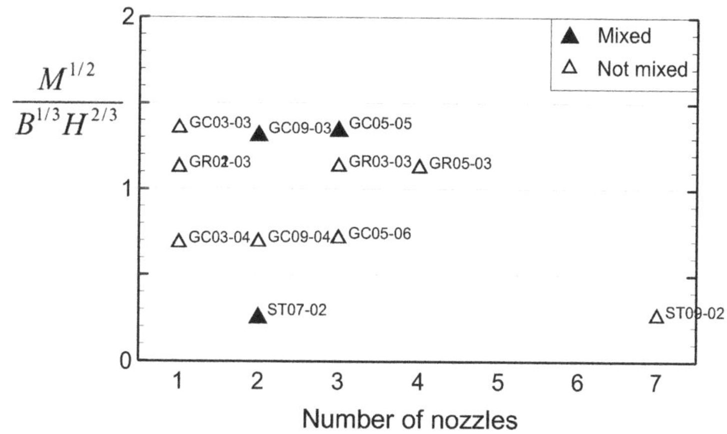

Figure 4.5 Mixing of negatively buoyant inflows for tanks with horizontal nozzles

Of the 12 tests with horizontal nozzles, only three resulted in mixing. They were the tanks with multiple nozzles that were elevated above the tank bottom. Even for those cases, however, configurations GC05 (3 nozzles at 1/3 and 2/3 depth) and GC09 (2 nozzles at mid depth), only mixed at high momentum fluxes, but not at lower momentum fluxes. Figure 4.5 indicates that the division between mixed and not mixed appears to be $M^{1/2}/B^{1/3}M^{2/3} \approx 1$. The standpipe tanks are an exception to this. ST07 (two nozzles, at mid depth and near the water surface) became mixed for $M^{1/2}/B^{1/3}M^{2/3} \approx 0.27$. Note, however, that ST09 (9 nozzles distributed over depth) did not become mixed for the same value of the momentum parameter.

Finally, as previously discussed, the draft tube can aid mixing. Tests ST did not become mixed for $(M\sin\theta)^{1/2}/B^{1/3}M^{2/3} = 0.36$ but did mix for the two tests with $(M\sin\theta)^{1/2}/B^{1/3}M^{2/3} \geq 0.33$. With no draft tube, the criterion was expected to be $(M\sin\theta)^{1/2}/B^{1/3}M^{2/3} \geq 0.85$ according to Eq. 4.3. In other words, the presence of the draft tube reduces the value of the critical momentum flux parameter above which mixing occurs.

In conclusion, negatively buoyant inflows mix best with either vertical nozzles at the bottom, or horizontal nozzles near the water surface. The momentum flux of the vertical bottom nozzle is opposite to the direction of gravity, and aids in raising the center of gravity of the tank water. The horizontal near-surface discharge uses the density difference to aid mixing as the buoyant jet falls to the bottom.

Positive buoyancy

For positively buoyant discharges (usually caused by warmer water entering colder water in the tank), conclusions opposite to the above would be expected. The experiments with positively buoyant inflows were run only with ground level cylindrical (GC tanks). The experiments with bottom nozzles are plotted in Figure 4.6.

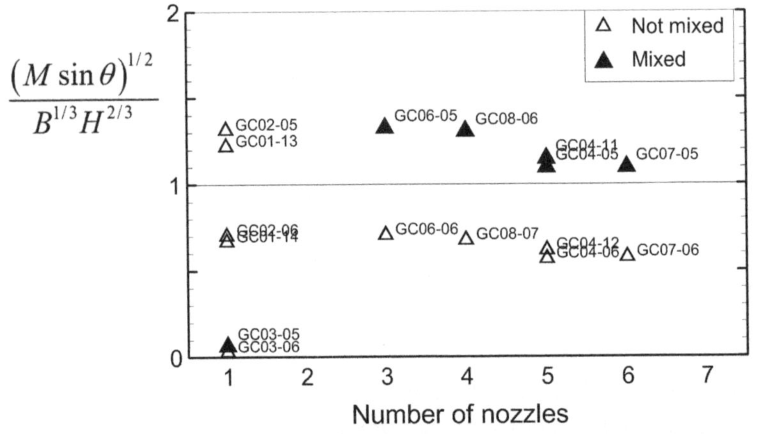

Figure 4.6 Mixing of positively buoyant inflows for all tanks with bottom inlets

Horizontal should be best. None of the single vertical nozzle configurations resulted in mixing. The only ones with upwardly inclined nozzles that did mix were those with multiple nozzles (three or more, no two nozzle configurations were tested) with high momentum flux. The tests with three or more nozzles that became mixed were those with $(M\sin\theta)^{1/2}/B^{1/3}M^{2/3}$ greater than about one. Lower values of this parameter did not mix. The single horizontal inlet became mixed for high momentum but not for low momentum.

To investigate this phenomenon further, the tests with horizontal nozzles were replotted in Figure 4.7 using the total momentum flux parameter. This clearly separates the results; the tests with $M^{1/2}/B^{1/3}H^{2/3} \geq 1.3$ became mixed, and those with $M^{1/2}/B^{1/3}H^{2/3} \leq 0.8$ did not.

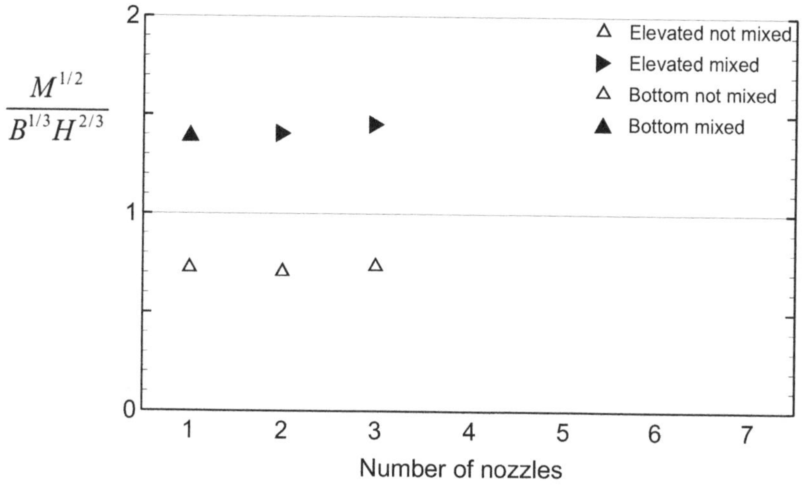

Figure 4.7 Mixing of positively buoyant inflows for all tanks with horizontal nozzles

CHAPTER 5.
CONCLUSIONS

Extensive experiments on the mixing induced by jets flowing into water storage tanks were conducted. The experiments involved scale models of cylindrical and square tank with dimensions that were typical of actual tanks. The complete distributions of tracer concentrations in the tanks were measured by a newly-developed three-dimensional laser-induced florescence (3DLIF) system that enabled accurate measurements of the extent of mixing in the tanks. Single and multiple-nozzle configurations were tested over a range of inflow velocities and density differences.

Nozzle design recommendations for various tanks were presented. Mixing can usually be accomplished by relatively simple nozzle configurations, provided they are suitably configured, and overly elaborate mixing devices are probably unnecessary. Density differences between the inflowing water and the water already in the tank impede mixing, but can be overcome by suitable nozzle orientation and adequate momentum flux. Criteria for mixing and recommendations are given for various tank shapes and density differences.

The experiments revealed complex flows that would be challenging to reliably simulate with present CFD (computational fluid dynamics) tools. The 3DLIF technique is a cost-effective technique for investigating mixing in water storage tanks.

APPENDIX A: APPLICATION OF RESULTS TO PROTOTYPE CONDITIONS

Introduction

The experiments described in this report were designed around the nominal prototype conditions (dimensions, flowrates and density differences) expected to be typical of operating storage tanks. The results can be scaled to other prototype conditions, however, by use of similitude. The means to do this, for flows with and without buoyancy effects, are discussed in this appendix.

a) No Buoyancy Effects

If the density of the inflow is the same as the density of the stored water, there are no buoyancy, or gravity effects on the flow. For this case, the dimensionless mixing time τ_m for cylindrical tanks is given by Eq. 2.7:

$$\tau_m = \frac{t_m M^{1/2}}{V^{2/3}} = f\left(\frac{H}{D}\right) \tag{A.1}$$

where t_m is the time for complete mixing (as defined in Chapter 3). Eq. A.1 assumes the nozzle diameter is small in comparison to H and D, i.e. it is only important inasmuch as it affects the momentum flux of the inflow. This is a valid approximation for most practical cases. The dimensionless time therefore depends on the ratio of the water depth to tank diameter and some data on this dependence were given in Figure 3.25. For the GC tanks, with $H/D = 0.25, \tau_m \approx 10$, and for ST tanks, with $H/D = 2.5$, $\tau_m \approx 15$. These values are for a single vertical nozzle discharging near the tank wall; τ_m differs for other inlet configurations as discussed in the report.

Consider, for example, a tank with height, $H = 30$ ft, diameter $D = 120$ ft, and an inflow rate, Q = 500 gpm = 1.11 ft^3/s, through a nozzle of diameter $d = 16$ in. The volume, $V = (\pi D^2/4)H = 339{,}000$ ft^3, the inflow velocity, $u_j = Q/(\pi/4)d^2$, and the momentum flux, $M = u_j Q = 0.88$ ft^4/s^2. The time for mixing t_m is:

$$t_m = \frac{\tau_m V^{2/3}}{M^{1/2}} \tag{A.2}$$

For this case, we have, $H/D = 0.25$, therefore $\tau_m \approx 10$. Solving Eq. A.2 yields $t_m = 52{,}000$ sec = 14.4 hours. Results for this and other flowrates are summarized in Table A.1.

Table A.1 Mixing Times for Prototype Cylindrical Tanks.

Diameter	Water depth	Inlet pipe diameter	Volume		Inflow rate		Jet velocity	Momentum flux	Mixing time Dimens-ionless	
D	H	d	V		Q		u_j	M	m	t_m
ft	ft	in	MG	ft^3	gpm	ft^3/s	ft/s	ft^4/s^2	-	hours
120	30	16	2.5	339,292	500	1.114	0.80	0.89	10	14.3
120	30	8	2.5	339,292	500	1.114	3.19	3.56	10	7.2
120	30	16	2.5	339,292	1000	2.228	1.60	3.56	10	7.2

For fixed tank volume, Eq. A.2 can be written:

$$t_m \propto \frac{\tau_m d}{Q} \qquad \text{(A.3)}$$

From which it can be seen that decreasing the nozzle diameter or increasing the flowrate decreases the mixing time in direction proportion to the changes in diameter or flowrate. This is evident in Table A.1, where a change in nozzle diameter from 16 to 8 inches or a doubling of the flowrate halves the mixing time. Of course, the penalty for the more rapid mixing is an increase in the pumping head and power required.

b) With Buoyancy Effects

With no buoyancy effects, Eq. A.3 shows that the results can be scaled to any flowrate. With density differences, however, they cannot be, and the scaling is more complicated.

With density differences, from Eq. 2.5:

$$\tau_m = \frac{t_m M^{1/2}}{D^2} = f\left(\frac{H}{D}, \frac{H}{l_M}\right) \qquad \text{(A.4)}$$

Where $l_M = M^{3/4} / B^{1/2}$ is a length-scale that expresses the distance over which the jet momentum is important relative to the jet buoyancy flux. Again, this equation assumes the nozzle diameter is small in comparison to H and D.

Similitude with buoyancy effects only applies when both the dimensionless ratios on the right hand side of Eq. A.4 are equal in model and prototype. (This is the same as the modeling laws, Eqs. 2.9 or 2.10) For example, experiment GC01-13 corresponds to a tank diameter D_1 = 119 ft, water depth H_1 = 30 ft, temperature difference of -10°F (positive buoyancy, the inflow is warmer than the water in the tank), and a flowrate Q_1 = 5,010 gpm through a nozzle of diameter d_1 = 16 inches (Table B.2). The tank did not become mixed for these conditions. Suppose another tank has a diameter D_2 = 80 ft, a flowrate Q_2 = 1,000 gpm and a nozzle of diameter d_2 = 12 inches. To what prototype temperature difference would this correspond?

For similarity, H/D must be the same, therefore the water depth of the second tank $H_2 = H_1 \times D_2 / D_1 = 30 \times 80/119 = 20.1$ ft. The ratio H/l_M must also be the same, so $l_{M2} = l_{M1} \times H_2 / H_1$.

For Tank 1, ΔT = -10°F, which is to a density difference $\Delta\rho \approx 1.0$ σ_t-units. Therefore:

Q_1 = 5010 gpm = 11.2 ft^3/s
u_{j1} = 8.0 ft/s
$M_1 = u_jQ$ = 89.4 ft^4/s^2
$B_1 = g(\Delta\rho/\rho_a)Q$ = 0.114 ft^4/s^3
$l_{M1} = M_1^{3/4}/B_1^{1/2}$ = 86.1 ft

For Tank 2:

Q_2 = 1000 gpm = 2.23 ft^3/s
u_{j2} = 2.84 ft/s
$M_2 = u_jQ$ = 6.32 ft^4/s^2
$l_{M2} = l_{M1}H_2/H_1$ = 57.7 ft = $M_2^{3/4}/B_2^{1/2}$ $\therefore B_2 = 0.0021$ ft^4/s^3 = $g(\Delta\rho/\rho_a)_2 Q_2$

Therefore, $\Delta\rho$ = 0.22 σ_t-units, which is a temperature difference of about 2.0°F.

The conditions for Tank 2 are then dynamically similar to those of Tank 1, and it can be concluded that a temperature difference of only 2.0°F would prevent the tank from mixing for these conditions. If it did mix, the time for mixing could be obtained from the experimentally determined value of τ_m in a manner similar to the calculation for the case above with no density difference.

For situations with density differences, two conditions for similitude must be satisfied. If the source momentum flux is arbitrarily chosen, the density difference for similitude must be computed as in the example above. Conversely, if the density difference is specified, then the source momentum flux must be computed. The source momentum flux and the density difference cannot both be arbitrarily specified.

APPENDIX B:
SUMMARY OF EXPERIMENTS

Table B.1 Summary of GC Tanks Experiments

(* For the experiments with no density difference, the inflow can be scaled to any prototype value)

Test	Prototype conditions									Model conditions												Results	
	Total no. of ports	Port diam.	Tank diam.	Water depth	Density diff.	Volume		Nominal Inflow*		Scale	Density diff. ratio	Port diam.	Tank diam.	Water depth	Density diff.	Volume	Inflow		Jet Reyn. No.	Jet Froude no.	H/D	Mixing time	
								Velcty	Rate								Velcty	Rate					
	n_p	d_{jp}	D	H	$\Delta\rho$	V		u_j	Q_T			d_j	D	H	$\Delta\rho$	V	u_j	Q_T	Re	F_j		t_m	τ_m
		in	ft	ft	σ_t	ft^3	MG	ft/s	gpm			in	in	in	σ_t	ft^3	ft/s	lpm				sec	
	GC01 (One port, bottom, side, vertical)																						
GC01-01	1	16	119	30	0	333,661	2.5	4.5	2,841	81.6	-	0.196	17.5	4.4	0.0	0.61	2.25	0.80	3,033	-	0.25	-	-
GC01-02	1	16	119	30	0	333,661	2.5	4.4	2,770	81.6	-	0.196	17.5	4.4	0.0	0.61	2.19	0.78	2,957	-	0.25	-	-
GC01-03	1	16	119	30	0	333,661	2.5	3.7	2,308	81.6	-	0.196	17.5	4.4	0.0	0.61	1.82	0.65	2,464	-	0.25	-	-
GC01-04	1	16	119	30	0	333,661	2.5	3.5	2,166	81.6	-	0.196	17.5	4.4	0.0	0.61	1.71	0.61	2,313	-	0.25	-	-
GC01-05	1	16	119	30	0	333,661	2.5	5.6	3,515	81.6	-	0.196	17.5	4.4	0.0	0.61	2.78	0.99	3,753	-	0.25	-	-
GC01-06	1	16	119	30	0	333,661	2.5	14.3	8,983	81.6	-	0.196	17.5	4.4	0.0	0.61	7.10	2.53	9,591	-	0.25	79.2	11.3
GC01-07	1	16	119	30	0	333,661	2.5	4.5	2,841	81.6	-	0.196	17.5	4.4	0.0	0.61	2.25	0.80	3,033	-	0.25	202.4	9.1
GC01-08	1	16	119	30	0	333,661	2.5	7.8	4,864	81.6	-	0.196	17.5	4.4	0.0	0.61	3.85	1.37	5,194	-	0.25	125.8	9.7
GC01-09	1	16	119	30	0	333,661	2.5	15.4	9,658	81.6	-	0.196	17.5	4.4	0.0	0.61	7.64	2.72	10,312	-	0.25	76.8	11.8
GC01-10	1	16	119	30	0	333,661	2.5	5.7	3,551	81.6	-	0.196	17.5	4.4	0.0	0.61	2.81	1.00	3,791	-	0.25	154.0	8.7
GC01-11	1	16	119	30	-1.15	333,661	2.5	7.5	4,687	81.6	0.05	0.196	17.5	4.4	-23.0	0.61	3.71	1.32	5,004	33.7	0.25	100.8	7.5
GC01-12	1	16	119	30	-1.15	333,661	2.5	2.8	1,775	81.6	0.05	0.196	17.5	4.4	-23.0	0.61	1.40	0.50	1,896	12.8	0.25	-	-
GC01-13	1	16	119	30	1.15	333,661	2.5	7.1	4,474	81.6	0.05	0.196	17.5	4.4	23.0	0.61	3.54	1.26	4,777	32.2	0.25	-	-
GC01-14	1	16	119	30	1.15	333,661	2.5	2.9	1,811	81.6	0.05	0.196	17.5	4.4	23.0	0.61	1.43	0.51	1,933	13.0	0.25	-	-
	GC01 (One port, bottom, side, vertical. Inflow and outflow at same rate)																						
GC01-15	1	16	119	30	0	333,661	2.5	11.3	7,101	81.6	-	0.196	17.5	4.4	0	0.61	5.61	2.00	7,582	-	0.25	120.0	13.5
GC01-16	1	16	119	30	0	333,661	2.5	6.1	3,799	81.6	-	0.196	17.5	4.4	0	0.61	3.00	1.07	4,056	-	0.25	228.0	13.7
GC01-17	1	16	119	30	0	333,661	2.5	6.0	3,764	81.6	-	0.196	17.5	4.4	0	0.61	2.98	1.06	4,018	-	0.25	-	-
GC01-18	1	16	119	30	0	333,661	2.5	6.1	3,799	81.6	-	0.196	17.5	4.4	0	0.61	3.00	1.07	4,056	-	0.25	218.4	13.1
	GC01 (One port, bottom, side, vertical. Variable H/D ratio)																						
GC01-19	1	16	119	60	0	667,321	5.0	6.1	3,835	81.6	-	0.196	17.5	8.8	0	1.23	3.03	1.08	4,094	-	0.50	221.2	8.5
GC01-20	1	16	119	60	0	667,321	5.0	15.0	9,409	81.6	-	0.196	17.5	8.8	0	1.23	7.44	2.65	10,046	-	0.50	122.0	11.5
GC01-21	1	16	119	120	0	1,334,643	10.0	14.9	9,338	81.6	-	0.196	17.5	17.6	0	2.46	7.38	2.63	9,970	-	1.01	189.6	11.1
GC01-22	1	16	119	90	0	1,000,982	7.5	12.2	7,634	81.6	-	0.196	17.5	13.2	0	1.84	6.04	2.15	8,151	-	0.76	165.6	9.6
	GC02 (One port, bottom, center, vertical)																						
GC02-01	1	16	119	30	0	333,661	2.5	15.4	9,658	81.6	-	0.196	17.5	4.4	0.0	0.61	7.64	2.72	10,312	-	0.25	73.6	11.3
GC02-02	1	16	119	30	0	333,661	2.5	5.7	3,551	81.6	-	0.196	17.5	4.4	0.0	0.61	2.81	1.00	3,791	-	0.25	364.0	20.5
GC02-03	1	16	119	30	-1.15	333,661	2.5	7.8	4,900	81.6	0.05	0.196	17.5	4.4	-23.0	0.61	3.87	1.38	5,232	35.2	0.25	103.2	8.0
GC02-04	1	16	119	30	-1.15	333,661	2.5	2.9	1,846	81.6	0.05	0.196	17.5	4.4	-23.0	0.61	1.46	0.52	1,971	13.3	0.25	-	-
GC02-05	1	16	119	30	1.15	333,661	2.5	8.0	5,006	81.6	0.05	0.196	17.5	4.4	23.0	0.61	3.96	1.41	5,345	36.0	0.25	-	-
GC02-06	1	16	119	30	1.15	333,661	2.5	3.1	1,953	81.6	0.05	0.196	17.5	4.4	23.0	0.61	1.54	0.55	2,085	14.0	0.25	-	-
GC02-07	1	16	119	30	0	333,661	2.5	14.3	8,983	81.6	-	0.196	17.5	4.4	0.0	0.61	7.10	2.53	9,591	-	0.25	96.0	13.7
GC02-08	1	16	119	30	0	333,661	2.5	12.2	7,670	81.6	-	0.196	17.5	4.4	0.0	0.61	6.06	2.16	8,189	-	0.25	117.6	14.3
GC02-09	1	16	119	30	0	333,661	2.5	10.2	6,391	81.6	-	0.196	17.5	4.4	0.0	0.61	5.05	1.80	6,824	-	0.25	154.0	15.6
GC02-10	1	16	119	30	0	333,661	2.5	8.1	5,078	81.6	-	0.196	17.5	4.4	0.0	0.61	4.01	1.43	5,421	-	0.25	201.6	16.2

Table B.1 Summary of GC Tanks Experiments

(* For the experiments with no density difference, the inflow can be scaled to any prototype value)

Test	Prototype conditions									Model conditions												Results	
	Total no. of ports	Port diam.	Tank diam.	Water depth	Density diff.	Volume		Nominal Inflow*		Scale	Density diff. ratio	Port diam.	Tank diam.	Water depth	Density diff.	Volume	Inflow		Jet Reyn. No.	Jet Froude no.	H/D	Mixing time	
								Velcty	Rate								Velcty	Rate					
	n_p	d_{jp}	D	H	$\Delta\rho$	V		u_j	Q_T			d_j	D	H	$\Delta\rho$	V	u_j	Q_T	Re	F_j		t_m	τ_m
		in	ft	ft	σ_t	ft^3	MG	ft/s	gpm			in	in	in	σ_t	ft^3	ft/s	lpm				sec	
	GC02 (One port, bottom, center, vertical. Inflow and outflow at same rate)																						
GC02-11	1	16	119	30	0	333,661	2.5	11.3	7,101	81.6	-	0.196	17.5	4.4	0	0.61	5.61	2.00	7,582	-	0.25	126.4	14.2
GC02-12	1	16	119	30	0	333,661	2.5	6.1	3,799	81.6	-	0.196	17.5	4.4	0	0.61	3.00	1.07	4,056	-	0.25	237.6	14.3
GC02-13	1	16	119	30	0	333,661	2.5	6.0	3,764	81.6	-	0.196	17.5	4.4	0	0.61	2.98	1.06	4,018	-	0.25	201.6	12.0
GC02-14	1	16	119	30	0	333,661	2.5	12.1	7,599	81.6	-	0.196	17.5	4.4	0	0.61	6.01	2.14	8,113	-	0.25	116.8	14.1
	GC03 (One port, bottom, side, horizontal)																						
GC03-01	1	16	119	30	0	333,661	2.5	16.4	10,262	81.6	-	0.196	17.5	4.4	0.0	0.61	8.11	2.89	10,956	-	0.25	73.6	12.0
GC03-02	1	16	119	30	0	333,661	2.5	6.1	3,853	81.6	-	0.196	17.5	4.4	0.0	0.61	3.05	1.09	4,113	-	0.25	179.2	10.9
GC03-03	1	16	119	30	-1.15	333,661	2.5	8.3	5,220	81.6	0.05	0.196	17.5	4.4	-23.0	0.61	4.13	1.47	5,573	37.5	0.25	-	-
GC03-04	1	16	119	30	-1.15	333,661	2.5	3.0	1,882	81.6	0.05	0.196	17.5	4.4	-23.0	0.61	1.49	0.53	2,009	13.5	0.25	-	-
GC03-05	1	16	119	30	1.15	333,661	2.5	8.0	5,042	81.6	0.05	0.196	17.5	4.4	23.0	0.61	3.99	1.42	5,383	36.3	0.25	120.0	9.6
GC03-06	1	16	119	30	1.15	333,661	2.5	3.1	1,917	81.6	0.05	0.196	17.5	4.4	23.0	0.61	1.52	0.54	2,047	13.8	0.25	-	-
	GC04 (Five ports arranged on two inlets pipes)																						
GC04-01	5	7.1	119	30	0	333,661	2.5	14.7	9,090	81.6	-	0.088	17.5	4.4	0.0	0.61	7.14	2.56	4,325	-	0.25	80.0	11.5
GC04-02	5	7.1	119	30	0	333,661	2.5	6.3	3,870	81.6	-	0.088	17.5	4.4	0.0	0.61	3.04	1.09	1,841	-	0.25	187.6	11.5
GC04-03	5	7.1	119	30	-1.15	333,661	2.5	8.3	5,149	81.6	0.05	0.088	17.5	4.4	-23.0	0.61	4.04	1.45	2,450	54.9	0.25	112.8	9.2
GC04-04	5	7.1	119	30	-1.15	333,661	2.5	3.1	1,882	81.6	0.05	0.088	17.5	4.4	-23.0	0.61	1.48	0.53	895	20.1	0.25	323.0	9.6
GC04-05	5	7.1	119	30	1.15	333,661	2.5	8.2	5,078	81.6	0.05	0.088	17.5	4.4	23.0	0.61	3.99	1.43	2,416	54.1	0.25	163.2	13.1
GC04-06	5	7.1	119	30	1.15	333,661	2.5	3.0	1,846	81.6	0.05	0.088	17.5	4.4	23.0	0.61	1.45	0.52	878	19.7	0.25	-	-
GC04-07	5	7.1	119	30	0	333,661	2.5	13.9	8,557	81.6	-	0.088	17.5	4.4	0.0	0.61	6.72	2.41	4,071	-	0.25	80.0	10.8
GC04-08	5	7.1	119	30	0	333,661	2.5	6.3	3,906	81.6	-	0.088	17.5	4.4	0.0	0.61	3.07	1.10	1,858	-	0.25	179.2	11.1
GC04-09	5	7.1	119	30	-1.15	333,661	2.5	8.2	5,078	81.6	0.05	0.088	17.5	4.4	-23.0	0.61	3.99	1.43	2,416	54.1	0.25	110.4	8.9
GC04-10	5	7.1	119	30	-1.15	333,661	2.5	3.1	1,882	81.6	0.05	0.088	17.5	4.4	-23.0	0.61	1.48	0.53	895	20.1	0.25	-	-
GC04-11	5	7.1	119	30	1.15	333,661	2.5	8.2	5,078	81.6	0.05	0.088	17.5	4.4	23.0	0.61	3.99	1.43	2,416	54.1	0.25	139.2	11.2
GC04-12	5	7.1	119	30	1.15	333,661	2.5	3.0	1,846	81.6	0.05	0.088	17.5	4.4	23.0	0.61	1.45	0.52	878	19.7	0.25	-	-
	GC05 (Three ports on one inlet pipe)																						
GC05-01	3	9.3	119	30	0	333,661	2.5	14.5	9,232	81.6	-	0.110	17.5	4.4	0.0	0.61	7.73	2.60	5,857	-	0.25	62.4	9.4
GC05-02	3	9.3	119	30	0	333,661	2.5	6.0	3,835	81.6	-	0.110	17.5	4.4	0.0	0.61	3.21	1.08	2,433	-	0.25	142.8	8.9
GC05-03	3	9.3	119	30	0	333,661	2.5	14.3	9,090	81.6	-	0.110	17.5	4.4	0.0	0.61	7.61	2.56	5,767	-	0.25	60.8	9.0
GC05-04	3	9.3	119	30	0	333,661	2.5	6.0	3,835	81.6	-	0.110	17.5	4.4	0.0	0.61	3.21	1.08	2,433	-	0.25	148.4	9.3
GC05-05	3	9.3	119	30	-1.15	333,661	2.5	7.9	5,006	81.6	0.05	0.110	17.5	4.4	-23.0	0.61	4.19	1.41	3,176	50.9	0.25	103.2	8.4
GC05-06	3	9.3	119	30	-1.15	333,661	2.5	3.0	1,917	81.6	0.05	0.110	17.5	4.4	-23.0	0.61	1.61	0.54	1,216	19.5	0.25	-	-
GC05-07	3	9.3	119	30	1.15	333,661	2.5	8.1	5,113	81.6	0.05	0.110	17.5	4.4	23.0	0.61	4.28	1.44	3,244	52.0	0.25	103.2	8.6
GC05-08	3	9.3	119	30	1.15	333,661	2.5	3.0	1,882	81.6	0.05	0.110	17.5	4.4	23.0	0.61	1.58	0.53	1,194	19.1	0.25	-	-

Table B.1 Summary of GC Tanks Experiments

(* For the experiments with no density difference, the inflow can be scaled to any prototype value)

Test	Prototype conditions									Model conditions												Results	
	Total no. of ports	Port diam.	Tank diam.	Water depth	Density diff.	Volume		Nominal Inflow*		Scale	Density diff. ratio	Port diam.	Tank diam.	Water depth	Density diff.	Volume	Inflow		Jet Reyn. No.	Jet Froude no.	H/D	Mixing time	
								Velcty	Rate								Velcty	Rate					
	n_p	d_{jp}	D	H	$\Delta\rho$	V		u_j	Q_T			d_j	D	H	$\Delta\rho$	V	u_j	Q_T	Re	F_j		t_m	τ_m
		in	ft	ft	σ_t	ft^3	MG	ft/s	gpm			in	in	in	σ_t	ft^3	ft/s	lpm				sec	
	GC06 (Three ports, centerline equally spaced, vertical)																						
GC06-01	3	9.3	119	30	0	333,661	2.5	14.5	9,196	81.6	-	0.110	17.5	4.4	0.0	0.61	7.70	2.59	5,834	-	0.25	51.2	7.7
GC06-02	3	9.3	119	30	0	333,661	2.5	6.0	3,835	81.6	-	0.110	17.5	4.4	0.0	0.61	3.21	1.08	2,433	-	0.25	140.0	8.8
GC06-03	3	9.3	119	30	-1.15	333,661	2.5	7.9	5,042	81.6	0.05	0.110	17.5	4.4	-23.0	0.61	4.22	1.42	3,199	51.3	0.25	88.8	7.3
GC06-04	3	9.3	119	30	-1.15	333,661	2.5	3.0	1,917	81.6	0.05	0.110	17.5	4.4	-23.0	0.61	1.61	0.54	1,216	19.5	0.25	248.2	7.8
GC06-05	3	9.3	119	30	1.15	333,661	2.5	7.8	4,935	81.6	0.05	0.110	17.5	4.4	23.0	0.61	4.13	1.39	3,131	50.2	0.25	96.0	7.7
GC06-06	3	9.3	119	30	1.15	333,661	2.5	3.0	1,882	81.6	0.05	0.110	17.5	4.4	23.0	0.61	1.58	0.53	1,194	19.1	0.25	-	-
	GC07 (Six ports on three inlet pipes)																						
GC07-01	6	6.5	119	30	0	333,661	2.5	14.5	8,983	81.6	-	0.080	17.5	4.4	0.0	0.61	7.11	2.53	3,918	-	0.25	60.8	8.7
GC07-02	6	6.5	119	30	0	333,661	2.5	6.2	3,835	81.6	-	0.080	17.5	4.4	0.0	0.61	3.04	1.08	1,673	-	0.25	131.6	8.0
GC07-03	6	6.5	119	30	-1.15	333,661	2.5	8.1	5,006	81.6	0.05	0.080	17.5	4.4	-23.0	0.61	3.96	1.41	2,184	56.4	0.25	96.0	7.6
GC07-04	6	6.5	119	30	-1.15	333,661	2.5	3.1	1,953	81.6	0.05	0.080	17.5	4.4	-23.0	0.61	1.55	0.55	852	22.0	0.25	282.2	8.7
GC07-05	6	6.5	119	30	1.15	333,661	2.5	8.1	5,042	81.6	0.05	0.080	17.5	4.4	23.0	0.61	3.99	1.42	2,199	56.8	0.25	153.6	12.3
GC07-06	6	6.5	119	30	1.15	333,661	2.5	3.0	1,882	81.6	0.05	0.080	17.5	4.4	23.0	0.61	1.49	0.53	821	21.2	0.25	-	-
	GC08 (Four ports, each quadrant, vertical)																						
GC08-01	4	8	119	30	0	333,661	2.5	15.3	9,587	81.6	-	0.098	17.5	4.4	0.0	0.61	7.58	2.70	5,118	-	0.25	-	-
GC08-02	4	8	119	30	0	333,661	2.5	6.1	3,835	81.6	-	0.098	17.5	4.4	0.0	0.61	3.03	1.08	2,047	-	0.25	100.8	6.1
GC08-03	4	8	119	30	0	333,661	2.5	14.6	9,161	81.6	-	0.098	17.5	4.4	0.0	0.61	7.24	2.58	4,890	-	0.25	43.2	6.3
GC08-04	4	8	119	30	-1.15	333,661	2.5	7.9	4,935	81.6	0.05	0.098	17.5	4.4	-23.0	0.61	3.90	1.39	2,635	50.2	0.25	86.4	6.8
GC08-05	4	8	119	30	-1.15	333,661	2.5	3.1	1,953	81.6	0.05	0.098	17.5	4.4	-23.0	0.61	1.54	0.55	1,043	19.9	0.25	258.4	8.0
GC08-06	4	8	119	30	1.15	333,661	2.5	8.0	5,042	81.6	0.05	0.098	17.5	4.4	23.0	0.61	3.99	1.42	2,692	51.3	0.25	88.8	7.1
GC08-07	4	8	119	30	1.15	333,661	2.5	2.9	1,846	81.6	0.05	0.098	17.5	4.4	23.0	0.61	1.46	0.52	986	18.8	0.25	-	-
GC08-08	4	8	119	30	-1.15	333,661	2.5	3.1	1,953	81.6	0.05	0.098	17.5	4.4	-23.0	0.61	1.54	0.55	1,043	19.9	0.25	255.0	7.9
	GC09 (Two ports on one inlet pipe)																						
GC09-01	2	11.3	119	30	0	333,661	2.5	15.6	9,729	81.6	-	0.139	17.5	4.4	0.0	0.61	7.65	2.74	7,326	-	0.25	57.6	8.9
GC09-02	2	11.3	119	30	0	333,661	2.5	6.1	3,835	81.6	-	0.139	17.5	4.4	0.0	0.61	3.02	1.08	2,888	-	0.25	142.8	8.7
GC09-03	2	11.3	119	30	-1.15	333,661	2.5	8.1	5,078	81.6	0.05	0.139	17.5	4.4	-23.0	0.61	3.99	1.43	3,824	43.1	0.25	103.2	8.3
GC09-04	2	11.3	119	30	-1.15	333,661	2.5	3.1	1,917	81.6	0.05	0.139	17.5	4.4	-23.0	0.61	1.51	0.54	1,444	16.3	0.25	-	-
GC09-05	2	11.3	119	30	1.15	333,661	2.5	8.1	5,078	81.6	0.05	0.139	17.5	4.4	23.0	0.61	3.99	1.43	3,824	43.1	0.25	120.0	9.6
GC09-06	2	11.3	119	30	1.15	333,661	2.5	3.0	1,846	81.6	0.05	0.139	17.5	4.4	23.0	0.61	1.45	0.52	1,390	15.7	0.25	-	-

Table B.2 Summary of ST Tanks Experiments

(* For the experiments with no density difference, the inflow can be scaled to any prototype value)

Test	Prototype conditions									Model conditions													Results	
	Total no. of ports	Port diam.	Tank diam.	Water depth	Density diff.	Volume		Nominal Inflow*		Scale	Density diff. ratio	Port diam.	Tank diam.	Water depth	Density diff.	Volume	Inflow		Jet Reyn. No.	Jet Froude no.	H/D	$V^{2/3}/M^{1/2}$	Mixing time	
								Velcty	Rate								Velcty	Rate						
	n_p	d_{jp}	D	H	$\Delta\rho$	V		u_j	Q_T			d_j	D	H	$\Delta\rho$	V	u_j	Q_T	Re	F_j			t_m	τ_m
		in	ft	ft	σ_t	ft³	MG	ft/s	gpm			in	in	in	σ_t	ft³	ft/s	lpm				sec	sec	
	ST01 (One port, bottom side, horizontal)																							
ST01-01	1	16.0	50.0	125	0	245,437	1.8	5.9	3,683	80.0	-	0.200	7.5	18.8	0	0.48	2.9	1.09	4,051	-	2.5	14.1	262.0	18.6
ST01-02	1	16.0	50.0	125	0	245,437	1.8	13.9	8,718	80.0	-	0.200	7.5	18.8	0	0.48	7.0	2.58	9,589	-	2.5	6.0	107.8	18.1
	ST02 (One port, bottom, side, vertical)																							
ST02-01	1	16.0	50.0	125	0	245,437	1.8	5.9	3,683	80.0	-	0.200	7.5	18.8	0	0.48	2.9	1.09	4,051	-	2.5	14.1	224.0	15.9
ST02-02	1	16.0	50.0	125	0	245,437	1.8	11.9	7,434	80.0	-	0.200	7.5	18.8	0	0.48	5.9	2.20	8,177	-	2.5	7.0	103.6	14.8
ST02-03	1	16.0	50.0	125	-1.15	245,437	1.8	3.0	1,859	80.0	0.05	0.200	7.5	18.8	-23.0	0.48	1.5	0.55	2,044	13.4	2.5	27.9	-	-
ST02-04	1	16.0	50.0	125	-0.54	245,437	1.8	3.1	1,926	80.0	0.05	0.200	7.5	18.8	-10.8	0.48	1.5	0.57	2,119	20.2	2.5	27.0	-	-
ST02-05	1	16.0	50.0	125	-0.26	245,437	1.8	3.0	1,859	80.0	0.05	0.200	7.5	18.8	-5.2	0.48	1.5	0.55	2,044	28.1	2.5	27.9	-	-
	ST02 (One port, bottom, side, vertical. Varing H/D)																							
ST02-06	1	16.0	50.0	100	0	196,350	1.5	14.7	9,225	80.0	-	0.200	7.5	15.0	0	0.38	7.4	2.73	10,147	-	2.0	4.9	70.0	14.4
ST02-07	1	16.0	50.0	100	0	196,350	1.5	6.3	3,954	80.0	-	0.200	7.5	15.0	0	0.38	3.2	1.17	4,349	-	2.0	11.3	156.0	13.8
ST02-08	1	16.0	50.0	75	0	147,262	1.1	6.0	3,751	80.0	-	0.200	7.5	11.3	0	0.29	3.0	1.11	4,126	-	1.5	9.9	120.0	12.2
ST02-09	1	16.0	50.0	75	0	147,262	1.1	15.0	9,428	80.0	-	0.200	7.5	11.3	0	0.29	7.5	2.79	10,370	-	1.5	3.9	49.0	12.5
ST02-10	1	16.0	50.0	50	0	98,175	0.7	6.0	3,785	80.0	-	0.200	7.5	7.5	0	0.19	3.0	1.12	4,163	-	1.0	7.5	70.0	9.4
ST02-11	1	16.0	50.0	50	0	98,175	0.7	15.3	9,597	80.0	-	0.200	7.5	7.5	0	0.19	7.7	2.84	10,555	-	1.0	2.9	28.8	9.8
ST02-12	1	16.0	50.0	25	0	49,087	0.4	6.1	3,818	80.0	-	0.200	7.5	3.8	0	0.10	3.0	1.13	4,200	-	0.5	4.7	46.2	9.9
ST02-13	1	16.0	50.0	25	0	49,087	0.4	11.3	7,096	80.0	-	0.200	7.5	3.8	0	0.10	5.7	2.10	7,805	-	0.5	2.5	25.2	10.1
	ST03 (One port, bottom, center, vertical)																							
ST03-01	1	16.0	50.0	125	0	245,437	1.8	13.8	8,617	80.0	-	0.200	7.5	18.8	0	0.48	6.9	2.55	9,478	-	2.5	6.0	91.0	15.1
ST03-02	1	16.0	50.0	125	0	245,437	1.8	5.8	3,616	80.0	-	0.200	7.5	18.8	0	0.48	2.9	1.07	3,977	-	2.5	14.4	218.0	15.2
ST03-03	1	16.0	50.0	125	-0.26	245,437	1.8	3.0	1,892	80.0	0.05	0.200	7.5	18.8	-5.2	0.48	1.5	0.56	2,081	28.6	2.5	27.4	-	-
	ST07 (Two ports, horizontal)																							
ST07-01	2	11.3	50.0	125	0	245,437	1.8	6.0	3,751	80.0	-	0.141	7.5	18.8	0	0.48	3.0	1.11	2,921	-	2.5	13.8	192.0	13.9
ST07-02	2	11.3	50.0	125	-1.15	245,437	1.8	3.0	1,892	80.0	0.05	0.141	7.5	18.8	-23.0	0.48	1.5	0.56	1,474	16.2	2.5	27.4	199.2	7.3
	ST09 (Seven ports, horizontal)																							
ST09-01	7	6.1	50.0	125	0	245,437	1.8	6.0	3,751	80.0	-	0.076	7.5	18.8	0	0.48	3.0	1.11	1,559	-	2.5	13.9	180.0	13.0
ST09-02	7	6.1	50.0	125	-1.15	245,437	1.8	3.0	1,859	80.0	0.05	0.076	7.5	18.8	-23.0	0.48	1.5	0.55	772	21.7	2.5	28.0	-	-
	ST11 (One port, center, vertical. With draft tube)																							
ST11-01	1	12.5	53.6	125	0	281,842	2.1	24.2	9,239	85.7	-	0.146	7.5	18.8	0	0.48	11.7	2.30	11,725	-	2.5	4.9	-	-
ST11-02	1	12.5	53.6	125	0	281,842	2.1	11.2	4,298	85.7	-	0.146	7.5	18.8	0	0.48	5.4	1.07	5,455	-	2.5	10.5	-	-
ST11-03	1	12.5	53.6	125	-1.15	281,842	2.1	5.8	2,209	85.7	0.05	0.146	7.5	18.8	-23.0	0.48	2.8	0.55	2,804	29.4	2.5	20.4	-	-
ST11-04	1	12.5	53.6	125	-0.54	281,842	2.1	5.8	2,209	85.7	0.05	0.146	7.5	18.8	-10.8	0.48	2.8	0.55	2,804	43.0	2.5	20.4	218.4	10.7
ST11-05	1	12.5	53.6	125	-0.26	281,842	2.1	5.9	2,249	85.7	0.05	0.146	7.5	18.8	-5.2	0.48	2.8	0.56	2,855	63.1	2.5	20.0	249.6	12.5
	ST12 (One port, center, vertical. No draft tube)																							
ST12-01	1	12.5	53.6	125	0	281,842	2.1	24.6	9,400	85.7	-	0.146	7.5	18.8	0	0.48	11.9	2.34	11,929	-	2.5	4.8	75.6	15.8
ST12-02	1	12.5	53.6	125	0	281,842	2.1	11.2	4,298	85.7	-	0.146	7.5	18.8	0	0.48	5.4	1.07	5,455	-	2.5	10.5	162.0	15.5
ST12-03	1	12.5	53.6	125	-1.15	281,842	2.1	5.9	2,249	85.7	0.05	0.146	7.5	18.8	-23.0	0.48	2.8	0.56	2,855	30.0	2.5	20.0	-	-
ST12-04	1	12.5	53.6	125	-0.26	281,842	2.1	5.9	2,249	85.7	0.05	0.146	7.5	18.8	-5.2	0.48	2.8	0.56	2,855	63.1	2.5	20.0	268.8	13.4
ST12-05	1	12.5	53.6	125	-0.54	281,842	2.1	5.9	2,249	85.7	0.05	0.146	7.5	18.8	-10.8	0.48	2.8	0.56	2,855	43.8	2.5	20.0	-	-

Table B.3 Summary of GR Tanks Experiments

(* For the experiments with no density difference, the inflow can be scaled to any prototype value)

Test	Prototype conditions									Model conditions													Results	
	Total no. of ports	Port diam.	Side wall length	Water depth	Density diff.	Volume		Nominal Inflow*		Scale	Density diff. ratio	Port diam.	Side wall length	Water depth	Density diff.	Volume	Inflow		Jet Reyn. No.	Jet Froude no.	H/D	$V^{2/3}/M^{1/2}$	Mixing time	
								Velcty	Rate								Velcty	Rate						
	n_p	d_{jp}	D	H	$\Delta\rho$	V		u_j	Q_T			d_j	D	H	$\Delta\rho$	V	u_j	Q_T	Re	F_j			t_m	τ_m
		in	ft	ft	σ_t	ft^3	MG	ft/s	gpm			in	in	in	σ_t	ft^3	ft/s	lpm				sec	sec	
	GR01 (One port, hottom horizontal, near floor, mid-wall)																							
GR01-01	1	16	105	30	0	330,750	2.5	6.1	3,835	81.6	-	0.196	15.4	4.4	0.0	0.609	3.03	1.08	4,093	-	0.29	16.4	151	9.2
GR01-02	1	16	105	30	0	330,750	2.5	15.8	9,908	81.6	-	0.196	15.4	4.4	0.0	0.609	7.83	2.79	10,575	-	0.29	6.3	67	10.6
GR01-03	1	16	105	30	-0.26	330,750	2.5	3.0	1,882	81.6	0.05	0.196	15.4	4.4	-5.2	0.609	1.49	0.53	2,009	-	0.29	33.3	-	-
	GR02 (One port, bottom horizontal, near the floor and wall)																							
GR02-01	1	16	105	30	0	330,750	2.5	6.0	3,764	81.6	-	0.196	15.4	4.4	0.0	0.609	2.98	1.06	4,018	-	0.29	16.7	158	9.5
GR02-02	1	16	105	30	0	330,750	2.5	15.9	9,944	81.6	-	0.196	15.4	4.4	0.0	0.609	7.86	2.80	10,613	-	0.29	6.3	62	9.9
GR02-03	1	16	105	30	-0.26	330,750	2.5	3.0	1,882	81.6	0.05	0.196	15.4	4.4	-5.2	0.609	1.49	0.53	2,009	28.5	0.29	33.3	-	-
	GR03 (Three ports, horizontal, mid-depth and near the wall)																							
GR03-01	3	9.3	105	30	0	330,750	2.5	6.1	3,871	81.6	-	0.114	15.4	4.4	0.0	0.609	3.02	1.09	2,369	-	0.29	16.3	120	7.4
GR03-02	3	9.3	105	30	0	330,750	2.5	15.3	9,731	81.6	-	0.114	15.4	4.4	0.0	0.609	7.59	2.74	5,956	-	0.29	6.5	48	7.4
GR03-03	3	9.3	105	30	-0.26	330,750	2.5	3.0	1,918	81.6	0.05	0.114	15.4	4.4	-5.2	0.609	1.50	0.54	1,174	37.5	0.29	32.9	-	-
	GR04 (One port,vertical, side)																							
GR04-01	1	16	105	30	0	330,750	2.5	6.1	3,835	81.6	-	0.196	15.4	4.4	0.0	0.609	3.03	1.08	4,093	-	0.29	16.4	142	8.7
GR04-02	1	16	105	30	0	330,750	2.5	15.9	9,944	81.6	-	0.196	15.4	4.4	0.0	0.609	7.86	2.80	10,613	-	0.29	6.3	98	15.5
GR04-03	1	16	105	30	-0.26	330,750	2.5	3.0	1,882	81.6	0.05	0.196	15.4	4.4	-5.2	0.609	1.49	0.53	2,009	28.5	0.29	33.3	260	7.8
GR04-04	1	16	105	30	0	330,750	2.5	15.9	9,944	81.6	-	0.196	15.4	4.4	0.0	0.609	7.86	2.80	10,613	-	0.29	6.3	93	14.7
GR04-05	1	16	105	30	0	330,750	2.5	15.6	9,766	81.6	-	0.196	15.4	4.4	0.0	0.609	7.72	2.75	10,423	-	0.29	6.4	99	15.4
GR04-06	1	16	105	30	-0.54	330,750	2.5	3.1	1,953	81.6	0.05	0.196	15.4	4.4	-10.8	0.609	1.54	0.55	2,085	20.5	0.29	32.1	246	7.7
GR04-07	1	16	105	30	-1.15	330,750	2.5	3.0	1,882	81.6	0.05	0.196	15.4	4.4	-23.0	0.609	1.49	0.53	2,009	13.5	0.29	33.3	-	-
	GR05 (Four ports at center and mid-depth)																							
GR05-01	4	8	105	30	0	330,750	2.5	6.1	3,835	81.6	-	0.098	15.4	4.4	0.0	0.609	3.03	1.08	2,047	-	0.29	16.4	106	6.5
GR05-02	4	8	105	30	0	330,750	2.5	15.1	9,482	81.6	-	0.098	15.4	4.4	0.0	0.609	7.49	2.67	5,060	-	0.29	6.6	38	5.8
GR05-03	4	8	105	30	-0.26	330,750	2.5	3.0	1,882	81.6	0.05	0.098	15.4	4.4	-5.2	0.609	1.49	0.53	1,004	40.2	0.29	33.3	-	-
GR05-02	4	8	105	30	0	330,750	2.5	15.2	9,518	81.6	-	0.098	15.4	4.4	0.0	0.609	7.52	2.68	5,079	-	0.29	6.6	37	5.6
	GR06 (Four ports on centerline, equally spaced, vertical)																							
GR06-01	4	8	105	30	0	330,750	2.5	6.1	3,835	81.6	-	0.098	15.4	4.4	0.0	0.609	3.03	1.08	2,047	-	0.29	16.4	168	10.3
GR06-02	4	8	105	30	0	330,750	2.5	15.2	9,553	81.6	-	0.098	15.4	4.4	0.0	0.609	7.55	2.69	5,098	-	0.29	6.6	62	9.5
GR06-03	4	8	105	30	-1.15	330,750	2.5	3.1	1,918	81.6	0.05	0.098	15.4	4.4	-23.0	0.609	1.52	0.54	1,023	19.5	0.29	32.7	246	7.5
GR06-04	4	8	105	30	0	330,750	2.5	6.1	3,835	81.6	-	0.098	15.4	4.4	0.0	0.609	3.03	1.08	2,047	-	0.29	16.4	166	10.1
GR06-05	4	8	105	30	0	330,750	2.5	14.7	9,234	81.6	-	0.098	15.4	4.4	0.0	0.609	7.30	2.60	4,927	-	0.29	6.8	64	9.4
	GR07 (Four ports on centerline, equally spaced, 45 degrees)																							
GR07-01	4	8	105	30	0	330,750	2.5	6.2	3,871	81.6	-	0.098	15.4	4.4	0.0	0.609	3.06	1.09	2,066	-	0.29	16.2	127	7.8
GR07-02	4	8	105	30	0	330,750	2.5	15.5	9,731	81.6	-	0.098	15.4	4.4	0.0	0.609	7.69	2.74	5,193	-	0.29	6.4	51	7.9
GR07-03	4	8	105	30	-0.26	330,750	2.5	3.0	1,882	81.6	0.05	0.098	15.4	4.4	-5.2	0.609	1.49	0.53	1,004	40.2	0.29	33.3	210	6.3
GR07-04	4	8	105	30	-0.54	330,750	2.5	3.1	1,918	81.6	0.05	0.098	15.4	4.4	-10.8	0.609	1.52	0.54	1,023	28.4	0.29	32.7	224	6.8
GR07-05	4	8	105	30	-1.15	330,750	2.5	3.1	1,953	81.6	0.05	0.098	15.4	4.4	-23.0	0.609	1.54	0.55	1,042	19.9	0.29	32.1	-	-
	GR08 (Four ports on side, equally spaced, 45 degrees)																							
GR08-01	4	8	105	30	0	330,750	2.5	6.1	3,835	81.6	-	0.098	15.4	4.4	0.0	0.609	3.03	1.08	2,047	-	0.29	16.4	158	9.7
[illegible]	4	8	105	30	0	330,750	2.5	15.3	9,589	81.6	-	0.098	15.4	4.4	0.0	0.609	7.58	2.70	5,117	-	0.29	6.5	61	9.3

REFERENCES

Daviero, G. J., Roberts, P. J. W., and Maile, K. (2001). "Refractive Index Matching in Large-Scale Stratified Experiments." Experiments in Fluids, 31, 119-126.

Duer, M.J., (2003). "Use of CFD to Analyze the Effects of Buoyant Inlet Jets on Mixing in Standpipes." *Proc. 2003 AWWA Annual Conference*. Anaheim, CA.

Fischer, H. B., List, E. J., Koh, R. C. Y., Imberger, J., and Brooks, N. H. (1979). "Mixing in Inland and Coastal Waters," Academic Press, New York.

Ferrier, A., Funk, D., and Roberts, P. J. W. (1993). "Application of Optical Techniques to the Study of Plumes in Stratified Fluids." Dynamics of Atmospheres and Oceans, 20, 155-183.

Fischer, H. B., List, E. J., Koh, R. C. Y., Imberger, J., and Brooks, N. H. (1979). "Mixing in Inland and Coastal Waters." Academic Press, New York.

Fossett, H., and Prosser, L. E. (1949). "The Application of Free Jets to the Mixing of Fluids in Bulk." *J. I. Mech. Eng*, 160, 224.

Grayman, W. M., Rossman, L. A., Arnold, C., Domineer, R. A., Smith, C., Smith, J. F., and Schnipke, R. (1999). "Water Quality Modeling of Distribution System Storage Facilities." Awwa Research Foundation, Denver, CO.

Lee, J. H. and Jirka, G. H. (1981), "Vertical round buoyant jet in shallow water." J. Hydr. Engrg., ASCE, 107(12), 1651-1675.

Martel, K. D., Martel, D., Kirmeyer, G. J., Murphy, B. M., Noran, P. F., and Kirby, L. (2002). "Preventing Water Quality Deterioration in Finished Water Storage Facilities." *Journal AWWA*, 94(4), 139-148.

Paik, J., Ge, L., and Sotiropoulos, F. (2004). "Recent Progress in Simulating Complex 3d Shear Flows Using Unsteady Statistical Turbulence Models." *International Journal of Heat and Fluid Flow*, 25(3), 513-527.

Paik, J., Sotiropoulos, F., and Sale, M. J. (2005). "Numerical Simulation of Swirling Flow in a Complex Hydro-Turbine Draft Tube Using Unsteady Statistical Turbulence Models." *Journal of Hydraulic Engineering*, In press, June 2005 issue.

Roberts, P. J. W., and Snyder, W. H. (1993). "Hydraulic Model Study for the Boston Outfall. I: Riser Configuration." Journal of Hydraulic Engineering, ASCE, 119(9), 970-987.

Roberts, P. J. W., and Toms, G. (1987). "Inclined Dense Jets in a Flowing Current." J. Hydr. Eng., ASCE, 113(3), 323-341.

Roberts, P. J. W. (2001). "Hydrodynamics of Initial Mixing Zones of Wastewater Discharges," Final Technical Report, EPA Grant Number R826216, School of Civil and Environmental Engineering, Georgia Institute of Technology, Atlanta, Georgia 30332.

Roberts, P. J. W., and Tian, X. (2000). "Three-Dimensional Imaging of Stratified Plume Flows." 5th International Conference on Stratified Flows, Vancouver, British Columbia, July 10 - 13, 2000.

Roberts, P. J. W., and Tian, X. (2002). "Application of Three-Dimensional Laser-Induced Fluorescence to Stratified Turbulent Mixing Processes." Hydraulic Measurements & Experimental Methods Conference 2002, Estes Park, Colorado, July 28 - August 1, 2002.

Rossman, L. A., and Grayman, W. M. (1999). "Scale-model studies of mixing in drinking water storage tanks." Journal of Environmental Engineering, ASCE, 125(8), 755-761.

Tian, X. (2002). "3DLIF and its Applications to Studies of the Near Field Mixing of Wastewater Discharges," PhD Thesis, School of Civil and Environmental Engineering, Georgia Institute of Technology, Atlanta, Ga.

Van de Vusse, J. G. (1955). "Mixing by Agitation of Miscible Liquids." *Chemical Engineering Science*, 4(5), 209-220.

Wright, S. J., Roberts, P. J. W., Zhongmin, Y., and Bradley, N. E. (1991). "Surface Dilution of Round Submerged Buoyant Jets." Journal of Hydraulic Research, 29(1), 67-89

ABBREVIATIONS

B	=	buoyancy flux of the inflow = $g_o'Q$
c	=	tracer (fluorescent dye) concentration
c_o	=	source tracer concentration in the inflow
COV	=	coefficient of variation
d	=	nozzle diameter
D	=	diameter of cylindrical tank
g	=	acceleration due to gravity
g_o'	=	modified acceleration due to gravity = $g_o' = g(\Delta\rho/\rho_a)$
$\mathrm{F_j}$	=	densimetric Froude number of inflow jet
H	=	water depth
l_M	=	length scale of jet with density difference = $M^{3/4}/B^{1/2}$
M	=	momentum flux of the inflow = u_jQ
n	=	total number or ports
Re	=	jet Reynolds number = u_jd/ν
Q	=	inflow rate
t	=	time
T_a	=	temperature of water in tank
T_o	=	temperature of inflow
V	=	volume of water in tank
u_j	=	inflow velocity
z_m	=	maximum rise height of a dense jet
ΔT	=	temperature difference between tank water and inflow = T_a - T_o
$\Delta\rho$	=	density difference between tank water and inflow = ρ_a - ρ_o
ν	=	kinematic viscosity
ρ_a	=	ambient density of water in tank
ρ_o	=	inflow density
τ	=	dimensionless time = $tM^{1/2}/V^{2/3}$
Θ	=	angle of nozzles to the horizontal

Subscripts:

m	=	model
p	=	prototype (full-size)
r	=	ratio of prototype to model, e.g. $D_r = D_p/D_m$

IWA Publishing's authorised EU representative for General Product Safety Regulations is Diane D'Arras, 15 rue Duret, 75116 Paris, France, e-mail: safety@iwap.co.uk.

Printed and bound by CPI Group (UK) Ltd, Croydon, CR0 4YY
17/09/2026
02218799-0011